Methods and Protocols in Food Science

Series Editor
Anderson S. Sant'Ana
University of Campinas
Campinas, Brazil

For further volumes:
http://www.springer.com/series/16556

Methods and Protocols in Food Science series is devoted to the publication of research protocols and methodologies in all fields of food science.

Volumes and chapters will be organized by field and presented in such way that the readers will be able to reproduce the experiments in a step-by-step style. Each protocol will be characterized by a brief introductory section, followed by a short aims section, in which the precise purpose of the protocol will be clarified.

Wine Analysis and Testing Techniques

Edited by

María Ángeles Pozo-Bayón and Carolina Muñoz González

CIAL, CSIC-UAM, Instituto de Investigación en Ciencias, Madrid, Madrid, Spain

Humana Press

Editors
María Ángeles Pozo-Bayón
CIAL, CSIC-UAM
Instituto de Investigación en Ciencias
Madrid, Madrid, Spain

Carolina Muñoz González
CIAL, CSIC-UAM
Instituto de Investigación en Ciencias
Madrid, Madrid, Spain

ISSN 2662-950X ISSN 2662-9518 (electronic)
Methods and Protocols in Food Science
ISBN 978-1-0716-3649-7 ISBN 978-1-0716-3650-3 (eBook)
https://doi.org/10.1007/978-1-0716-3650-3

This Humana imprint is published by the registered company Springer Science+Business Media, LLC, part of Springer Nature.
The registered company address is: 1 New York Plaza, New York, NY 10004, U.S.A.

Paper in this product is recyclable.

Preface to the Series

Methods and Protocols in Food Science series is devoted to the publication of research protocols and methodologies in all fields of food science. The series is unique as it includes protocols developed, validated, and used by food and related scientists as well as theoretical basis provided for each protocol. Aspects related to improvements in the protocols, adaptations, and further developments in the protocols may also be approached.

Methods and Protocols in Food Science series aims to bring the most recent developments in research protocols in the field as well as very well-established methods. As such, the series targets undergraduate, graduate, and researchers in the field of food science and correlated areas. The protocols documented in the series will be highly useful for scientific inquiries in the field of food sciences, presented in such way that the readers will be able to reproduce the experiments in a step-by-step style.

Each protocol will be characterized by a brief introductory section, followed by a short aims section, in which the precise purpose of the protocol is clarified. Then, an in-depth list of materials and reagents required for employing the protocol is presented, followed by a comprehensive and step-by-step procedures on how to perform that experiment. The next section brings the dos and don'ts when carrying out the protocol, followed by the main pitfalls faced and how to troubleshoot them. Finally, template results will be presented and their meaning/conclusions addressed.

The Methods and Protocols in Food Science series will fill an important gap, addressing a common complaint of food scientists, regarding the difficulties in repeating experiments detailed in scientific papers. With this, the series has a potential to become a reference material in food science laboratories of research centers and universities throughout the world.

Preface

Wine is a globally consumed and appreciated high-quality beverage with a very complex chemical composition achieved through a very careful process involving the selection of the vine and grape varieties and the use of appropriate viticulture and winemaking techniques. A large array of many different types of chemical compounds and their correct proportion determine its quality. Therefore, their analysis through the application of robust and rigorous procedures is essential to prove wine safety, quality, and authenticity.

Some rapid and classical "wet chemistry" methods commonly used for wine analysis include measurement of pH, titratable and volatile acidity, soluble solids, sulfur dioxide, ethanol, and molecular spectroscopy methods (total phenolic, anthocyanins, color, α-amino acids, and enzymatic-coupled reactions). Many of these methods are used for routine analysis in many labs and even in the own cellar. Most of these procedures are compiled as standard methods for wine analysis by different international organizations (e.g., International Organisation of Vine and Wine). Additionally, many indirect spectroscopy techniques and sensor-based methods are being gaining in popularity to quantify different types of chemical compounds. To do so, commercial instruments are currently available.

However, for a precise wine chemical analysis, more sensitive and specific instrumental methods are necessary. These techniques will become even more relevant in the upcoming years since the oenological sector will have to face new challenges associated to climate change (introduction of new grape varieties, adaptation of fermentative processes and new microorganisms, new oenological additives), reduction in the use of preservatives, and/or the verification of wine's varietal authenticity and the confirmation of geographic origin. These are only some examples of the toughest challenges for wine analytical chemists and control laboratories. In this sense, they will require very precise cutting-edge analytical procedures and techniques to ensure wine safety, quality, and authenticity. Therefore, techniques based in mass spectrometry and new holistic approaches (metabolomics, volatilomics) are very sensitive and promising tools for wine testing.

Moreover, in recent years, the application of sensory sciences has proved to be essential to understand the sensory meaning of the chemical changes in sensory perception and consumer response. In this sense, together with the application of the well-established descriptive sensory analysis, new rapid techniques for sensory profiling and hedonic/preference and emotional tests are providing crucial information in product development. In addition to this, well-known (time-intensity) and more recent methods (temporal dominance sensations) that allow us to monitor the evolution of sensory perception through wine tasting combined with new in vivo analytical tools are becoming more and more used in order to understand the relationship between wine chemistry and the sensory experience during wine tasting.

The objective of this book is to provide to the reader a wide spectrum of instrumental and sensory methods that currently used for wine testing. The first part of the book (Chaps. 1–9) is focused on instrumental methods, while the second part of the book is aimed to provide a selection of sensory procedures (Chaps. 10–16), and Chap. 17 is related to the application of gas chromatography-olfactometry combining both approaches. Selected well-validated and published procedures are described by experts on the different topics, providing an introduction, including a specific application in wine, listing all the materials and reagents and comprehensively describing all the procedures step-by-step. Additionally, a large number of notes giving tips on troubleshooting and for avoiding known pitfalls are included.

This book will intend to compile the state of the art in instrumental and sensory wine testing procedures for a broad range of wine applications. This, will be useful for young researchers (i.e., graduate students, postdoctoral) and all researchers from industry or academia who are either still at the begin of their academic career or senior scientists who are in search for new challenges in a new field hitherto unfamiliar to them.

Coordinated by María Ángeles Pozo-Bayón and Carolina Muñoz González from the Spanish National Council of Research (CSIC), this book brings together the collaboration of many recognized scientists around the world, to whom we would like to thank their valuable contribution for making possible this book.

Madrid, Spain

María Ángeles Pozo-Bayón
Carolina Muñoz González

Contents

Contributors

CRISTINA ALCALDE-EON • *Facultad de Farmacia, Grupo de Investigación en Polifenoles, Departamento de Química Analítica, Nutrición y Bromatología, Universidad de Salamanca, Salamanca, Spain*

JACQUES EMMANUEL BARBIER • *Institut Œnologique de Champagne, Epernay, France*

CRISTINA V. BERENGUER • *CQM – Centro de Química da Madeira, Natural Products Research Group, Universidade da Madeira, Funchal, Portugal*

PAOLO BOCCACCI • *Institute for Sustainable Plant Protection, National Research Council (IPSP-CNR), Torino, Italy*

JEANNE BRAND • *South African Grape and Wine Research Institute, Department of Viticulture and Oenology, Stellenbosch University, Stellenbosch, South Africa*

JOSÉ SOUSA CÂMARA • *CQM – Centro de Química da Madeira, Natural Products Research Group, Universidade da Madeira, Funchal, Portugal; Departamento de Química, Faculdade de Ciências Exatas e Engenharia, Universidade da Madeira, Funchal, Portugal*

CAROLINA CHAYA • *Department of Agricultural Economics, Statistics, and Business Management, Universidad Politécnica de Madrid, Escuela Técnica Superior de Ingeniería Agronómica, Alimentaria y de Biosistemas, Madrid, Spain*

CELIA CRIADO • *Instituto de Investigación en Ciencias de la Alimentación (CIAL) CSIC-UAM, Madrid, Spain*

MARÍA TERESA ESCRIBANO-BAILÓN • *Facultad de Farmacia, Grupo de Investigación en Polifenoles, Departamento de Química Analítica, Nutrición y Bromatología, Universidad de Salamanca, Salamanca, Spain*

GIORGIO GAMBINO • *Institute for Sustainable Plant Protection, National Research Council (IPSP-CNR), Torino, Italy*

IGNACIO GARCÍA-ESTÉVEZ • *Facultad de Farmacia, Grupo de Investigación en Polifenoles, Departamento de Química Analítica, Nutrición y Bromatología, Universidad de Salamanca, Salamanca, Spain*

ATTILA GERE • *Institute of Food Science and Technology, Department of Postharvest, Supply Chain, Commerce and Sensory Science, Hungarian University of Agriculture and Life Sciences, Budapest, Hungary*

ADAM M. GILMORE • *HORIBA Instruments Inc, Piscataway, NJ, USA*

GAMZE GUCLU • *Department of Food Engineering, Faculty of Engineering, Cukurova University, Adana, Türkiye*

KATE HOWELL • *School of Agriculture, Food and Ecosystem Sciences, Faculty of Science, The University of Melbourne, Melbourne, VIC, Australia*

DAVID W. JEFFERY • *School of Agriculture, Food and Wine, and Waite Research Institute, The University of Adelaide, Glen Osmond, SA, Australia*

HASIM KELEBEK • *Department of Food Engineering, Faculty of Engineering, Adana Alparslan Turkes Science and Technology University, Adana, Türkiye*

OZLEM KILIC-BUYUKKURT • *Department of Food Technology, Faculty of Kadirli Applied Sciences, Osmaniye Korkut Ata University, Osmaniye, Türkiye*

ZOLTÁN KÓKAI • *Institute of Food Science and Technology, Department of Postharvest, Supply Chain, Commerce and Sensory Science, Hungarian University of Agriculture and Life Sciences, Budapest, Hungary*

Jean-Luc Legras • *SPO, INRAE, Institut Agro, University Montpellier, Montpellier, France*

Di Liu • *State Key Laboratory of Nutrient Use and Management, College of Resources and Environmental Sciences, China Agricultural University, Beijing, China*

Richard Marchal • *Laboratoire d'Œnologie, Université de Reims Champagne-Ardenne, Reims – Université de Haute-Alsace, Colmar, France*

Matthew McSweeney • *School of Nutrition and Dietetics, Acadia University, Wolfville, NS, Canada*

María Mora • *BCC Innovation, Technology Center in Gastronomy, Basque Culinary Center, Donostia-San Sebastián, Spain; Basque Culinary Center, Faculty of Gastronomic Sciences, Mondragon Unibersitatea, Donostia-San Sebastián, Spain*

Carolina Muñoz González • *Instituto de Investigación en Ciencias de la Alimentación (CIAL) CSIC-UAM, Madrid, Spain*

José M. Oliveira • *CEB – Centre of Biological Engineering, University of Minho, Braga, Portugal; LABBELS – Associate Laboratory, Braga/Guimarães, Portugal*

Jorge A. M. Pereira • *CQM – Centro de Química da Madeira, Natural Products Research Group, Universidade da Madeira, Funchal, Portugal*

Rosa Perestrelo • *CQM – Centro de Química da Madeira, Natural Products Research Group, Universidade da Madeira, Funchal, Portugal*

María Pérez-Jiménez • *Mediterranean Institute for Agricultural, Environmental and Development (MED), University of Évora, Évora, Portugal*

María Ángeles Pozo-Bayón • *Instituto de Investigación en Ciencias de la Alimentación (CIAL) CSIC-UAM, Madrid, Spain*

Ranaweera K. R. Ranaweera • *School of Agriculture, Food and Wine, and Waite Research Institute, The University of Adelaide, Glen Osmond, SA, Australia*

Bertrand Robillard • *Institut Œnologique de Champagne, Epernay, France*

Juan José Rodríguez-Bencomo • *Department of Analytical Chemistry, Faculty of Chemical Science, Complutense University of Madrid, Madrid, Spain*

Maria Pilar Sáenz-Navajas • *Department of Enology, Instituto de Ciencias de la Vid y del Vino (UR-CSIC-GR), Logroño, Spain*

Erika Salas • *Facultad de Farmacia, Grupo de Investigación en Polifenoles, Departamento de Química Analítica, Nutrición y Bromatología, Universidad de Salamanca, Salamanca, Spain; Facultad de Ciencias Químicas, Universidad Autónoma de Chihuahua, Chihuahua, México*

Thomas Salmon • *Laboratoire d'Œnologie, Université de Reims Champagne-Ardenne, Reims - Université de Haute-Alsace, Colmar, France*

Serkan Selli • *Department of Food Engineering, Faculty of Engineering, Cukurova University, Adana, Türkiye*

Onur Sevindik • *Department of Food Engineering, Faculty of Engineering, Adana Alparslan Turkes Science and Technology University, Adana, Türkiye*

I. Nyoman Sumerta • *School of Agriculture, Food and Ecosystem Sciences, Faculty of Science, The University of Melbourne, Melbourne, VIC, Australia; National Research and Innovation Agency (BRIN), Jakarta, Indonesia*

Dominique Valentin • *Centre des Sciences du Goût et de l'Alimentation, CNRS, INRAE, Institut Agro, Université de Bourgogne, Dijon, France*

MAR VILANOVA • *Instituto de Ciencias de la Vid y del Vino - ICVV, Spanish National Research Council-CSIC, University of La Rioja and Government of La Rioja, Logroño, Spain*

ALICE VILELA • *Chemistry Research Center (CQ-VR), Department of Agronomy, School of Agrarian and Veterinary Sciences, University of Trás-os-Montes and Alto Douro, Vila Real, Portugal*

Part I

Instrumental Methods

Chapter 1

DNA-Based Methods for Wine Traceability and Varietal Authentication Using Single Nucleotide Polymorphism Genotyping Assays

Paolo Boccacci and Giorgio Gambino

Abstract

The final characteristics of a wine are strongly influenced by must varietal composition. Further, wine quality and value can be heavily modified if grape varieties other than those expected/allowed are used, especially in the case of monovarietal wines. Thus, genetic traceability is an important tool used to protect the authenticity of commercial wines against fraud. Single nucleotide polymorphisms (SNPs) are a promising molecular marker class to reach this aim in wines. In this chapter, we have reported a DNA extraction protocol and SNP TaqMan® genotyping assays for varietal identification and wine authentication, developed for *Vitis vinifera* L. 'Nebbiolo' musts and wines.

Key words DNA extraction, Genetic traceability, Grapevine, Quantitative polymerase chain reaction (qPCR), SNPs, TaqMan® probe, Wine

1 Introduction

The vinification process and the geographical origin of the grapes are two important parameters of wine quality. Nevertheless, the final characteristics of the wine are strongly influenced by must varietal composition, especially in monovarietal wines, for which only one cultivar is used. Thus, wine authenticity and market price can be heavily modified if grape varieties other than those allowed by the production regulations are employed.

Traditional approaches used for the varietal identification and authenticity of musts and wines are based on methods that associate chemical composition with variety and production area [1–3]. Nevertheless, chemometric approaches are often expensive in terms of time and resources, and they present some inaccuracies caused by the great influence of viticultural and winemaking methods on the qualitative and quantitative composition of wine [1]. Therefore,

María Ángeles Pozo-Bayón and Carolina Muñoz González (eds.), *Wine Analysis and Testing Techniques*, Methods and Protocols in Food Science, https://doi.org/10.1007/978-1-0716-3650-3_1,

the results cannot be considered reliable if the models are applied to commercial wines [4].

DNA typing has proved to be a valuable technique for accurately identifying cultivars due to its independence from external conditions and its high discriminating power. By extracting DNA from must and wine and using molecular markers amplified by polymerase chain reaction (PCR), it is possible to identify the grape variety used in winemaking. However, the results can be very different depending on the wines, the DNA extraction technique, the type of marker, and the amplification technique used. Among the available DNA markers, microsatellites or simple-sequence repeats (SSRs) represent the most common markers used in grapevine for fingerprinting [5]. Thus, nuclear and chloroplast SSR markers have been used to identify cultivars, analyzing residual grape DNA extracted from mono-varietal and multi-varietal musts and wines [6–15]. All authors obtained positive results in must analysis but reported reproducibility problems for the systematic authentication of finished experimental and/or commercial wines. The main limiting factors were the low DNA quality and quantity, mainly due to DNA degradation during the winemaking processes, reduction of DNA quantity by clarification and filtration of wines, presence of yeasts' DNA, and PCR inhibitors, such as polyphenols, polysaccharides, and proteins.

Single-nucleotide polymorphisms (SNPs) are considered the newest type of molecular marker for grapevine identification. They are particularly interesting because are most abundant in the genome and have the potential to become a valid alternative to SSRs [16]. Moreover, SNPs can be employed to overcome the degradation limitations, allowing DNA amplification using more sensitive techniques, such as quantitative real-time polymerase chain reaction (qPCR). Although SNP information content is lower compared with SSRs and many SNP makers are required for varietal identification, several genome sequencing or re-sequencing of different grape cultivars, such as *Vitis vinifera* L. cv 'Nebbiolo' [17], have allowed the identification of several mutations and polymorphisms between different genotypes. Therefore, SNPs have also been used for genetic traceability of varieties in wine [10, 18, 19].

In our protocol, the most effective approach is the analysis of a limited number of SNP markers using qPCR, which allows the identification of a specific cultivar within a group of genotypes [20, 21]. It was developed for the varietal identification and authentication of 'Nebbiolo'-based wines, such as Barolo and Barbaresco, using SNP TaqMan® genotyping assays, based on two SNP markers (SNP_15082 and SNP_14783) that were sufficient to distinguish 'Nebbiolo' from a group of 1157 true-to-type genotypes [20]. Moreover, the high sensitivity of the assays allowed identifying, for the first time, mixtures of 1% and 10% of 'Barbera' in 'Nebbiolo' musts at the end of maceration and of malolactic fermentation, respectively, as well as contamination of 10–20% of 'Barbera' in 'Nebbiolo' experimental wines [20].

2 Materials

2.1 DNA Extraction Protocol

1. Extraction buffer: 1 M Tris–HCl, 1.4 M NaCl, 20 mM EDTA (pH 8.0), 3% w/v cetyltrimethylammonium bromide (CTAB), autoclaved at 120 °C for 20 min and stored at room temperature.
2. β-mercaptoethanol, stored at 4 °C.
3. Chloroform–isoamyl alcohol (24:1 v/v), stored at room temperature.
4. 10% w/v cetyltrimethylammonium bromide (CTAB), stored at room temperature.
5. Ethanol, stored at −20 °C.
6. TE buffer: 10 mM Tris–HCl (pH 8.0) and 1 mM EDTA (pH 8.0), stored at room temperature.
7. Proteinase K (20 mg/mL), stored at −20 °C.
8. Phenol:chloroform:isoamyl alcohol (25:24:1), stored at 4 °C.
9. Ammonium acetate (7.5 M), stored at 4 °C.
10. 70% ethanol prepared with autoclaved ultrapure water and stored at −20 °C.
11. DNA extraction NucleoSpin® Plant Kit (Macherey-Nagel, Düren, Germany) (*see* **Note 1**).

2.2 SNP Genotyping Assay

1. TaqMan® Environmental Master Mix 2.0 (Thermo Fisher Scientific, Waltham, MA, USA) (*see* **Note 2**).
2. TaqMan® SNP Genotyping Assay mix (Thermo Fisher Scientific, Waltham, MA, USA), containing premixed forward and reverse primers, VIC probe, and FAM probe (Table 1) (*see* **Note 2**).
3. Sterile ultrapure water.

3 Methods

3.1 Wine Sample Preparation

1. Transfer the wine from the glass bottle into a plastic bottle and store at −20 °C (*see* **Note 3**).
2. After at least 2 days, thaw and homogenize vigorously the wine sample by shaking the bottle several times (*see* **Note 4**).

Table 1
Primers and probes specific for 'Nebbiolo' used for SNP genotyping

ID marker	Allele Nebbiolo	Allele non-Nebbiolo	ID oligo	Primer and probe sequences 5′-3′	Length of the fragment (bp)
SNP_14783	G	A	For	GAGCACAATCAACAATTATCCA TTT	83
			Rev	TGGTTGTGTTAATAGCAGGCAA	
			Probe Allele A	FAM-TAAAAAAGTGTTAAGGTGA TAAT-NFQ	
			Probe Allele G	VIC-TAAAAAAGTGTTAAGGT GATGAT-NFQ	
SNP_15082	T	C	For	TCTCTTCTGGCATGGAAATCAAT	89
			Rev	TAGATTACGGGCCAAGCTGA	
			Probe Allele T	FAM-TCTCATTTTCCTCATTAT-NFQ	
			Probe Allele C	VIC-TCTCATTTTCCTCATCATG-NFQ	

3. Collect 100 mL of wine in two separate 50 mL tubes.
4. Centrifuge at 4000 *g* for 1 h at 4 °C.
5. Discard the supernatant and continue with the DNA extraction or store the pellet at −20 °C until all the necessary samples have been prepared.

3.2 DNA Extraction

1. Preheat the water bath to 65 °C and precool centrifuge at 4 °C (*see* **Note 5**).
2. Prewarm at 65 °C the extraction buffer and add β-mercaptoethanol (10 μL for 1 mL of extraction buffer) before use (*see* **Note 6**).
3. Use 5 mL of prewarmed extraction buffer containing β-mercaptoethanol for each wine sample and dissolve the pellet by vortex (*see* **Note 7**).
4. Incubated at 65 °C for 1 h, with mixing every 10–15 min.
5. Add 5 mL (1 volume) of chloroform:isoamyl alcohol (24:1), vortex vigorously, and centrifuge at 8000 *g* for 10 min at 4 °C (*see* **Note 8**).
6. Transfer the aqueous phase (top layer) from each tube into a new 16 mL tube, taking care to leave behind the bottom layer and the debris-filled interface, and mix with 0.1 volume of 10% CTAB prewarmed at 65 °C (*see* **Note 9**).
7. Add 1 volume of chloroform:isoamyl alcohol (24:1), vortex vigorously, and centrifuge at 8000 *g* for 10 min at 4 °C.

8. Recover the aqueous phase (top layer) from each tube into a new 16 mL tube, taking care to leave behind the bottom layer and the debris-filled interface, and add 2 volumes of cold ethanol. Mix by inverting the tube and incubate at −25 °C overnight.
9. The next day, preheat the water bath to 48 °C and precool the centrifuge at 4 °C.
10. Centrifuge at 10,000 *g* for 30 min at 4 °C.
11. Eliminate supernatant and resuspend the pellet in 250 μL of TE buffer. Transfer each sample from the 16 mL tube to the 2 mL tube. Add 20 μL of Proteinase K (20 mg/mL) and incubate at 48 °C for 30 min.
12. Add 1 volume (270 μL) of phenol:chloroform:isoamyl alcohol (25:24:1), vortex vigorously, and centrifuge at 11,000 *g* for 15 min at 4 °C (*see* **Note 10**).
13. Recover the aqueous phase (top layer) from each tube into a new 2 mL tube, taking care to leave behind the bottom layer and the debris-filled interface (*see* **Note 11**). Add 2 volumes of cold ethanol and 2.5 M of ammonium acetate 7.5 M. Mix by inverting the tube and incubate at −25° for at least 2 h.
14. Centrifuge at 22,000 *g* for 30 min at 4 °C.
15. Eliminate supernatant, wash the pellet with 500 μL of 70% cold ethanol, and centrifuge at 22,000 *g* for 10 min at 4 °C.
16. Eliminate supernatant and dissolve the pellet in 100 μL of ultrapure sterile water.
17. Finally, add 65 μL of cold ethanol and 115 μL of Buffer PC from NucleoSpin® Plant Kit. Load samples onto the Nucleospin® Plant II Column (green ring) and process according to the manufacturer's instructions. Perform the final elution in 45 μL of Buffer PE and store at −20 °C (*see* **Note 12**).
18. Quantification of the DNA using a NanoDrop spectrophotometer (*see* **Note 13**).

3.3 SNP Genotyping Assay

1. Thaw the DNA samples and sterile ultrapure water at room temperature and thaw the qPCR reagents in ice. Vortex and spin.
2. For each wine sample, prepare a mix containing: 2.25 μL of sterile ultrapure water, 5 μL TaqMan® Environmental Master Mix 2.0 (*see* **Note 14**), 0.25 μL of 40× TaqMan® SNP Genotyping Assay, containing premixed forward and reverse primers, VIC probe and FAM probe (*see* **Note 15**). Vortex and spin.
3. Distribute 7.5 μL mix per reaction in a PCR plate or in 0.2 mL strip PCR tubes.

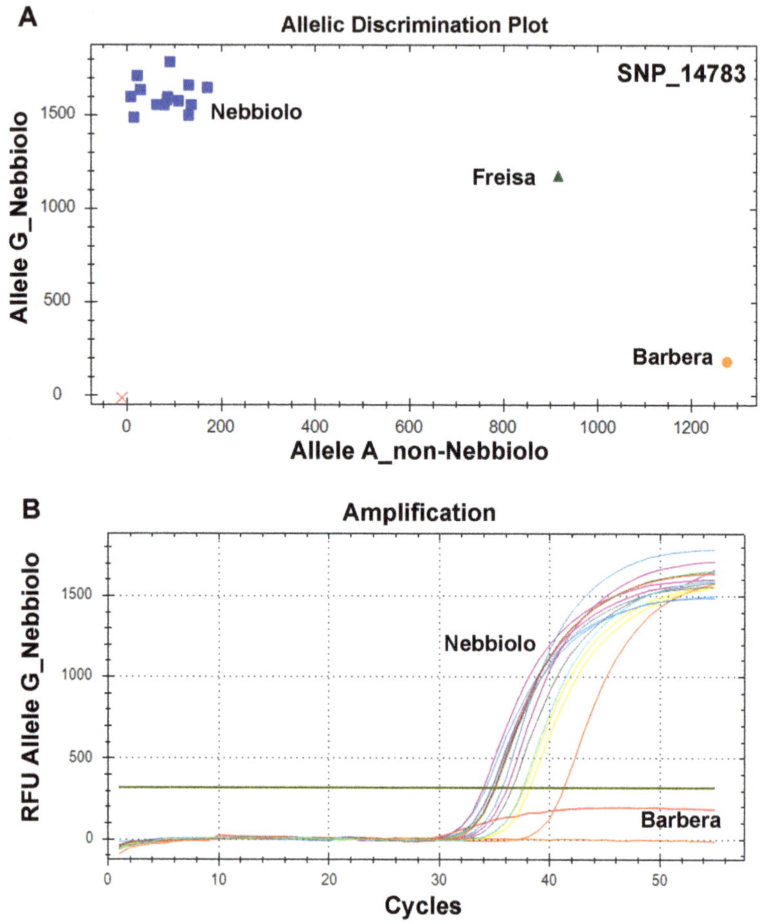

Fig. 1 SNP genotyping in 'Nebbiolo' wines. (**a**) Scatterplot of TaqMan® SNP_14783 genotyiping assay with 'Nebbiolo' wines. (**b**) Relative fluorescence unit (RFU) of the TaqMan® SNP_14783 probe tagged with VIC dye (allele G 'Nebbiolo'). The yellow line in the amplification plot indicates the RFU level of 'Barbera' (non-'Nebbiolo' control), above which, it was possible to detect 'Nebbiolo' wines. The control DNA from 'Nebbiolo,' 'Barbera' (homozygous non-'Nebbiolo' genotype), and 'Freisa' (heterozygous genotype) were extracted from leaves

4. Add 2.5 μL of DNA extracted from each wine sample (*see* **Note 16**). Spin plate or tubes using a plate centrifuge.
5. Run the PCR on a thermal cycler for real-time PCR with the following conditions: initial denaturation step at 95 °C for 10 min, followed by 55 cycles of 95 °C for 15 s, and 60 °C for 1 min.
6. Allelic discrimination plots (Fig. 1) are constructed following the instruction manual of the thermocycler manufacturer.

4 Notes

1. For final DNA purification, NucleoSpin® Plant Kit (Macherey-Nagel, Düren, Germany) is recommended due to the good quality of the DNA obtained. However, this kit could be replaced by another commercial kit for DNA extraction from plant tissue.
2. The TaqMan® Master Mix indicated is the enzyme that gives the best performance with the protocol reported. It is theoretically possible to use other TaqMan® Master Mix from different companies, but it will probably be necessary to re-calibrate the PCR reaction parameters. In addition, the indicated TaqMan® SNP Genotyping Assay is strongly recommended because these probes use Minor Groove Binder (MGB) technology and are particularly effective for discriminating between highly homologous allelic sequences. However, it would be possible to use other probes from different companies, but it will probably be necessary to recalibrate the PCR reaction.
3. Storage of wine samples at −20 °C contributes to initial precipitation of nucleic acids, combining the low temperature with the presence of alcohol in the wine, which occurs until complete sample freezing.
4. While thawing the wine sample, more homogeneous wine aliquots are obtained by shaking or inverting the bottle several times.
5. The DNA extraction protocol was previously reported [20, 21] and it was developed from the CTAB-based method reported by Siret et al. [22] and modified by Agrimonti and Marmiroli [12].
6. The step using β -mercaptoethanol should be performed under a fume hood.
7. DNA extraction starts from 100 mL for wine, centrifuged in two 50 mL tubes. Then, the 5 mL of prewarmed extraction buffer containing β-mercaptoethanol required should be divided equally (2.5 mL) into the two 50 mL tubes. After dissolving the pellet from each of the two 50 mL tubes in 2.5 mL of extraction buffer containing β-mercaptoethanol, transfer and combine everything in a single 16 mL tube.
8. Chloroform:isoamyl alcohol (24:1) and phenol:chloroform: isoamyl alcohol (25:24:1) should be used under a fume hood.
9. The high concentration of CTAB (100 g per L) used in this solution causes precipitations during storage. Therefore, the solution should be preheated to 65 °C until the CTAB has dissolved.

10. Phenol extraction is crucial for good DNA extraction from wine samples [23].
11. During the recovery of the aqueous phase, it is essential to avoid subsequent contamination by phenol. This reagent has a maximum absorbance at 270/275 nm, which is close to the DNA absorbance (260 nm). Phenol contamination can bias DNA quantification with higher yields and purity due to an upward shift in the absorbance value at 260 nm. Moreover, PCR reaction can be inhibited by the presence of phenol.
12. The final purification using a commercial DNA extraction kit helps in reducing the level of contaminants and PCR inhibitors, such as polyphenolic compounds, proteins, and polysaccharides.
13. The DNA quality and quantity is generally estimated using a NanoDrop spectrophotometer by determining the spectrophotometric absorbance of the samples at 230 nm (polysaccharides), 260 nm (nucleic acids and phenolic compounds), and 280 nm (proteins) and the ratios of $A_{260/A280}$ and $A_{260/A230}$. Nevertheless, several previous works [13, 24] reported the presence of yeast DNA and phenolic substances in the DNA extracted from the wine. Thus, the data obtained with Nanodrop quantification are overestimated and spectrophotometric quantification is often not reliable for the quantification of grapevine DNA in wine. Alternatively, can be adopted a more specific quantification of grapevine DNA-based on the qPCR amplification of the endogenous gene of *V. vinifera VvNCED2* using TaqMan® probes [6, 13, 20, 21, 24].
14. The TaqMan® Environmental Master Mix 2.0 (Thermo Fisher Scientific, Waltham, MA, USA) may be replaced by another TaqMan® Master Mix, from a different manufacturer, with good tolerance to PCR inhibitors.
15. In Table 1 was reported an example of TaqMan® SNP genotyping assays specific for 'Nebbiolo' wines [20, 21]. The protocol can be applied to other wines from different cultivars after the identification and validation of specific SNPs. In addition, for different wines, it may be necessary to slightly modify the DNA extraction protocol.
16. For each wine sample, analyze at least three replicates obtained from three different DNA extractions.

References

1. Versari A, Laurie VF, Ricci A et al (2014) Progress in authentication, typification and traceability of grapes and wines by chemometric approaches. Food Res Int 60:2–18. https://doi.org/10.1016/j.foodres.2014.02.007
2. Villano C, Lisanti MT, Gambuti A et al (2017) Wine varietal authentication based on phenolics, volatiles and DNA markers: state of the art, perspectives, and drawbacks. Food Control 80: 1–10. https://doi.org/10.1016/j.foodcont.2017.04.020
3. Solovyev PA, Fauhl-Hassek C, Riedl J et al (2021) NMR spectroscopy in wine authentication: an official control perspective. Compr Rev Food Sci Food Saf 20:2040–2062. https://doi.org/10.1111/1541-4337.12700
4. Zhang J, Li L, Gao N et al (2010) Feature extraction and selection from volatile compounds for analytical classification of Chinese red wines from different varieties. Anal Chim Acta 662:137–142. https://doi.org/10.1016/j.aca.2009.12.043
5. This P, Jung A, Boccacci P et al (2004) Development of a standard set of microsatellite reference alleles for identification of grape cultivars. Theor Appl Genet 109:1448–1458. https://doi.org/10.1007/s00122-004-1760-3
6. Bigliazzi J, Scali M, Paolucci E et al (2012) DNA extracted with optimized protocols can be genotyped to reconstruct the varietal composition of monovarietal wines. Am J Enol Vitic 63:568–573. https://doi.org/10.5344/ajev.2012.12014
7. Boccacci P, Akkak A, Torello Marinoni D et al (2012) Genetic traceability of Asti Spumante and Moscato d'Asti musts and wines using nuclear and chloroplast microsatellite markers. Eur Food Res Technol 235:439–446. https://doi.org/10.1007/s00217-012-1770-3
8. Pereira L, Martins-Lopes P, Batista C et al (2012) Molecular markers for assessing must varietal origin. Food Anal Methods 5:1252–1259. https://doi.org/10.1007/s12161-012-9369-7
9. Recupero M, Garino C, Paolis D et al (2013) A method to check and discover adulteration of Nebbiolo-based monovarietal musts: detection of Barbera and Dolcetto cv via SSR analysis coupled with Lab-On-Chip® microcapillary electrophoresis. Food Anal Methods 6:952–962. https://doi.org/10.1007/s12161-012-9506-3
10. Catalano V, Moreno-Sanz P, Lorenzi S et al (2016) Experimental review of DNA-based methods for wine traceability and development of a Single-Nucleotide Polymorphism (SNP) genotyping assay for quantitative varietal authentication. J Agric Food Chem 64:6969–6984. https://doi.org/10.1021/acs.jafc.6b02560
11. di Rienzo V, Miazzi MM, Fanelli V et al (2016) An enhanced analytical procedure to discover table grape DNA adulteration in industrial musts. Food Control 60:124–130. https://doi.org/10.1016/j.foodcont.2015.07.015
12. Agrimonti C, Marmiroli N (2018) PCR analysis of experimental and commercial wines by means of nuclear and chloroplast SSRs. Eur Food Res Technol 244:2127–2140. https://doi.org/10.1007/s00217-018-3121-5
13. Vignani R, Liò P, Scali M (2019) How to integrate wet lab and bioinformatics procedures for wine DNA admixture analysis and compositional profiling: case studies and perspectives. PLoS One 14:e0211962. https://doi.org/10.1371/journal.pone.0211962
14. Zambianchi S, Soffritti G, Stagnati L et al (2021) Applicability of DNA traceability along the entire wine production chain in the real case of a large Italian cooperative winery. Food Control 124:107929. https://doi.org/10.1016/j.foodcont.2021.107929
15. Zambianchi S, Soffritti G, Stagnati L et al (2022) Effect of storage time on wine DNA assessed by SSR analysis. Food Control 142: 109249. https://doi.org/10.1016/j.foodcont.2022.109249
16. Cabezas JA, Ibáñez J, Lijavetzky D et al (2011) A 48 SNP set for grapevine cultivar identification. BMC Plant Biol 11:153. https://doi.org/10.1186/1471-2229-11-153
17. Gambino G, Dal Molin A, Boccacci P et al (2017) Whole-genome sequencing and SNV genotyping of 'Nebbiolo' (*Vitis vinifera* L.) clones. Sci Rep 7:17294. https://doi.org/10.1038/s41598-017-17405-y
18. Pereira L, Gomes S, Castro C et al (2017) High Resolution Melting (HRM) applied to wine authenticity. Food Chem 216:80–86. https://doi.org/10.1016/j.foodchem.2016.07.185
19. Barrias S, Fernandes JR, Eiras-Dias JE (2019) Label free DNA-based optical biosensor as a potential system for wine authenticity. Food Chem 270:299–304. https://doi.org/10.1016/j.foodchem.2018.07.058
20. Boccacci P, Chitarra W, Schneider A et al (2020) Single nucleotide polymorphism (SNP) genotyping assays for the varietal authentication of 'Nebbiolo' musts and wines.

Food Chem 312:126100. https://doi.org/10.1016/j.foodchem.2019.126100

21. Gambino G, Ferrero L, Scalzini G et al (2022) Impact of oenological processing aids and additives on the genetic traceability of 'Nebbiolo' wine produced with withered grapes. Food Res Int 151:110874. https://doi.org/10.1016/j.foodres.2021.110874
22. Siret R, Gigaud O, Rosec JP et al (2002) Analysis of grape *Vitis vinifera* L. DNA in must mixtures and experimental mixed wines using microsatellite markers. J Agric Food Chem 50:3822–3827. https://doi.org/10.1021/jf011462e
23. Pereira L, Guedes-Pinto H, Martins-Lopes P (2011) An enhanced method for *Vitis vinifera* L. DNA extraction from wines. Am J Enol Vitic 62:547–552. https://doi.org/10.5344/ajev.2011.10022
24. Savazzini F, Martinelli L (2006) DNA analysis in wines: development of methods for enhanced extraction and real-time polymerase chain reaction quantification. Anal Chim Acta 563:274–282. https://doi.org/10.1016/j.aca.2005.10.078

Chapter 2

Optimization of Microbial DNA Extraction from Wine, Juice, and Sap for Community-Based Genome Studies

I. Nyoman Sumerta, Jean-Luc Legras, Kate Howell, and Di Liu

Abstract

Obtaining high-quality DNA for diversity analysis from plant-based samples is challenging due to the presence of inhibitors, such as phenolics which are naturally abundant in plant tissues. Phenolics can hinder DNA extraction processes causing degradation, enzymatic inhibition, and oxidation, leading to low yield and purity DNA for analysis of microbial communities. Modifying the extraction method by performing a pre-treatment step to reduce the concentration of phenolics will be crucial before the extraction process of plant-based samples. Here we describe the addition of polyvinylpolypyrrolidone (PVPP) to bind phenolics, which is then discarded from the extraction process. DNA extraction process is completed by following commercially available extraction kits for efficient release of high-quality DNA for downstream analyses. This method can be used for extracting DNA from plant-derived liquids (grape juice, must, wine; also xylem/phloem sap inside the plant) for genomic studies.

Key words DNA extraction, Genomics, Polyphenols, Polyvinylpolypyrrolidone, Grape, Wine, Xylem/phloem sap

1 Introduction

Assessing microbial communities directly from the environment has received huge attention in genomics. It contributes to a deeper insight of the ecological process that benefits food industries with applications widespread for agricultural, health, and environmental analyses [1–3]. Using advanced technology in genomics to reveal microbial communities improves the scalability and accuracy uncovered by traditional microbiological methods, which were largely based on the ability of the microorganism to be cultured in the laboratory [1, 4]. To ensure accurate representation, DNA extraction is the crucial stage to determine the quantity, quality, and validity of the microbial composition analyzed by microbiome sequencing like next-generation sequencing (NGS).

Obtaining good DNA quality for revealing the microbial community is challenging. Isolating sufficient yield of sequencing

María Ángeles Pozo-Bayón and Carolina Muñoz González (eds.), *Wine Analysis and Testing Techniques*, Methods and Protocols in Food Science, https://doi.org/10.1007/978-1-0716-3650-3_2,

quality DNA depends on the type of samples and the extraction method [5], and affects not only the quantity and quality of DNA but also downstream molecular analyses [6, 7]. Methods have been developed to obtain a good yield of DNA into extraction kits providing efficiency and reproducibility [8]. On the other hand, each type of sample has inhibitors that determine DNA yield quality. For instance, a sample from plant materials contains some metabolites that could interfere the DNA extraction, such as polysaccharides and phenolics compounds that should be removed to optimize the DNA yield and downstream analyses [8–10]. If those substances remain after extraction, they would inhibit enzymatic activities [8, 9]. For example, those inhibitors could prevent the binding of taq-polymerase during PCR and hinder the enzymatic modification of restriction endonuclease digestion [11]. The presence of phenolic substances also leads to DNA degradation [10].

In food and beverage production, the application of genomics data to reveal the microbial communities has provided deep information about the fermentation microecology [12]. The fermentation of plant products contains many inhibitor compounds; thus, it is challenging to obtain an appropriate genetic material. Different plant materials have inhibitors in various concentrations, such as polysaccharides and secondary metabolites which could affect the yielded DNA quality [13], and the choice of DNA extraction method would affect community diversity index values [5]. Samples of grape juice, must, wine fermentation and also xylem/phloem sap inside the plant (e.g., grapevine, palm) are rich in polysaccharides and phenolics which need to be removed to obtain good quality DNA for downstream molecular applications. With several strategies, cetyl-trimethyl ammonium bromide (CTAB) in buffer extraction can remove the polysaccharides [9, 14, 15], and the combination of sodium chloride, potassium acetate, and isopropanol can precipitate DNA from polysaccharides [13], 2-mercaptoethanol protects DNA from oxidation and degradation, and polyvinylpolypyrrolidone (PVPP) contributes to polyphenols removal [9, 16]. In samples with high content of polyphenols, PVPP can tightly bind to polyphenols and separate these complexes from non-phenolics compounds [16, 17].

DNA extraction kits provide time efficiency and allow modifications to improve the DNA yield that suits downstream applications. DNA extraction kits are used for investigating microbial communities in next-generation sequencing giving reproducibility between extractions and allowing extraction from a wider range of sample types [18]. A wide range of kits that could extract different types of samples is helpful for scalability samples in microbial community studies [5]. Our previously published works have combined the application of PVPP during sample preparation to remove phenolic compounds as an extraction step prior to a standard kit protocol and achieved good quality data [19, 20]. Therefore, we described the techniques used for wine, juice, and sap samples in the study of Liu et al. [19] in this optimization protocol. DNA was isolated from pellets after centrifugation, washed with PVPP for several times, extracted with commercial kit, and finally the quality and quantity of DNA were assessed using gel visualization, Nanodrop, and Qubit measurements.

2 Materials

Prepare all materials and store them at appropriate temperatures. All materials can be stored at room temperature, excepting electrophoresis dye and ladder at −20 °C and Qubit standards at 4 °C. Follow all instructions and waste management procedures for hazardous materials.

2.1 DNA Exaction Kit DNeasy PowerSoil Kit (Qiagen, Cat. No. 12888-100)

- This protocol uses DNeasy PowerSoil kit* as an example but also works for other commercial DNA exaction kits (please follow manufacturer's instructions accordingly), for example, FastDNA™ Kit.
- A little modification at the beginning before following the manufacturer's instructions (*see* Subheading 3.2).

2.2 0.1 M PBS (Phosphate-Buffered Saline), pH 5.2

- Take 80 mL of double distilled water in a volumetric flask.
- Add 2.0214 g of sodium phosphate dibasic heptahydrate to the solution.
- Add 0.3394 g of sodium phosphate monobasic monohydrate to the solution.
- Adjust solution to pH 5.2 using HCl or NaOH.
- Add distilled water until the volume is 100 mL.
- Sterilize by filtration using a 0.22 μm membrane.

2.3 1% PVPP (Polyvinylpolypyrrolidone) (Sigma-Aldrich, Cat. No. 81420-500G or ThermoFisher Scientific Cat. No. J62417)

- Add 1.0 g of PVPP to an appropriate flask and add 0.1 M PBS to 100 mL volume.
- Mix and store at 4 °C.

2.4 1× TAE (Tris-Acetate-EDTA) Buffer

- Prepare stock solution in 1.0 L of 10× TAE.
- Add 48.5 g of Tris to volumetric flask and dissolve with 800 mL of distilled water.
- Add 11.4 mL of acetic acid glacial.
- Add 20 mL of 0.5 M EDTA (pH 8.0).

- Mix thoroughly and add distilled water to 1.0 L.
- Store the stock solution at room temperature.
- For 1× TAE, dilute the stock solution at 1:10 factor.

2.5 SYBR™ Safe Gel Staining (Invitrogen, Cat. No. S33102 or GLPBIO Cat. No. GC12792)

- Environmentally safe and less hazardous gel stain for DNA compared to ethidium bromide and light sensitive (store in the dark container at room temperature).
- Alternatively, other similar gel stains are also available, for instance, GelRed® nucleic acid gel stain.

2.6 1.5% Agarose

- Prepare 0.75 g of Agarose to the flask and add 50 mL of 1× TAE solution.
- Mix thoroughly and microwave for 1 min.
- Wait until slightly warm (±50 °C) and add 5 μL of gel staining (SYBR safe).
- Shake roundly until agarose solution is stained thoroughly.
- Pour agarose solution into the electrophoresis tray and set the comb.
- Leave for 20 min at room temperature or 10 min at 4 °C fridge to harden the gel.

2.7 6× Loading Dye

- The loading dye is used when running the electrophoresis.
- Pipette mixing the DNA with loading dye before loading to the gel wells (*see* Subheading 3.3)

2.8 1 kb DNA Ladder

- For referencing the DNA quantity. Mostly the ladder is colorless and mixed with loading dye (2:1) before loading to the gel wells.

(b) Qubit buffers and standards (Invitrogen, Cat. No. Q33230 or Fisher Scientific Cat. No. 15880210)

  - There are two standards and working buffer to run the qubit measurement (*see* Subheading 3.4.2).

3 Methods

DNA isolation using a commercial kit, DNeasy PowerSoil, with some modifications in sample preparation before following manufacturer's instructions.

3.1 Sample Preparation

1. Thaw the frozen samples at 4 °C.
2. Centrifuge liquid samples of wine, juice, and sap at 10,000 × *g* for 10 min.
3. Remove supernatant or transfer to another tube for further analysis.

4. Wash pellets with the 500 μL pre-cooled 1% PVPP in 0.1 M PBS (*see* **Note 1**).
5. Vortex for 15 s.
6. Centrifuge the mixture at 10,000 × *g* for 10 min.
7. Remove the supernatant and return to **step 4** (pipette mixing with pellet if required).
8. Repeat **steps 4–6** for 2–3 times (*see* **Note 2**).
9. Store the pellet at −20 °C for the next analysis.

3.2 DNA Extraction

10. Transfer 200–300 μL of extraction buffer taken from the PowerBead Tube to the pellet, and pipette-mixing thoroughly. Then, retransfer to the PowerBead Tube (*see* **Note 3**).
11. Add 60 μL of Solution C1 to the PowerBead Tube and vortex for 10 s. If precipitated, heat solution C1 at 60 °C.
12. Vortex horizontally the PowerBead Tubes for 10 min or Bead beater in 2000 Hz for 2 min.
13. Centrifuge the PowerBead at 10,000 × *g* for 30 s.
14. Prepare a new 2 mL Collection Tube and transfer 400–500 μL of supernatant to prepared Collection Tube.
15. Add 250 μL of Solution C2 and vortex briefly. Put in chiller for 5 min incubation to optimize solution C2 precipitation to organic and inorganic contaminants such as protein and cell debris.
16. Centrifuge the collection tubes at 10,000 × *g* for 1 min.
17. Transfer up to 600 μL of supernatant to a new 2 mL Collection Tube. Avoid the pellet while pipetting.
18. Add 200 μL of Solution C3 and homogenize briefly. Put in chiller for 5 min incubation to optimize solution precipitation.
19. Centrifuge the collection tubes at 10,000 × *g* for 1 min.
20. Transfer 750 μL of supernatant to a new 2 mL Collection Tube, avoid the pellet while transferring.
21. Mix Solution C4 by shaking and add 1200 μL to the supernatant. Vortex briefly.
22. Add 675 μL mixed solution from **step 12** to MB Spin Column and centrifuge at 10,000 × *g* for 1 min. Discard flow-through. Repeat this step twice until all of mixed solution is processed.
23. Add 500 μL of Solution C5. Centrifuge at 10,000 × *g* for 30 s.
24. Discard the flow-through. Centrifuge again at 10,000 × *g* for 1 min.
25. Put MB Spin Column into a new 2 mL Collection Tube by avoiding any splashes of Solution C5 and dry it up for 5 min.

26. Add 50–100 μL of Solution C6 or alternatively used DNA-free PCR-grade water to the center of the white filter membrane of MB Spin Column. Ensure the entire membrane is wet to have better yield.
27. Centrifuge a at 10,000 × *g* for 30 s. Discard the MB spin column from collection tube. The eluted DNA is now ready for quality check and then store at −20 °C or −80 °C for further downstream applications.

3.3 Gel Imaging

28. Prepare gel agarose and 1× TAE solution (*see* Subheading 2.6).
29. Add 5 μL of SBYR™ safe to the warm gel mixture (*see* **Note 4**).
30. Pour the agarose on a gel tray with comb. Wait until the agar hardens.
31. Release the comb and put the gel to the electrophoresis chamber containing 1× TAE, make sure the gel soaks properly until no air bubbles in wells.
32. Prepare parafilm to mix 1 kb ladder and 6× loading dye. Load 2.5 μL of loading dye onto parafilm and add with 5 μL ladder or each sample. Pipette mixing thoroughly (*see* **Note 5**).
33. Load to the well in agarose gel and run to the electrophoresis apparatus for 60 min with 75 Voltage.
34. Visualize the DNA on GelDoc (Fig. 1).

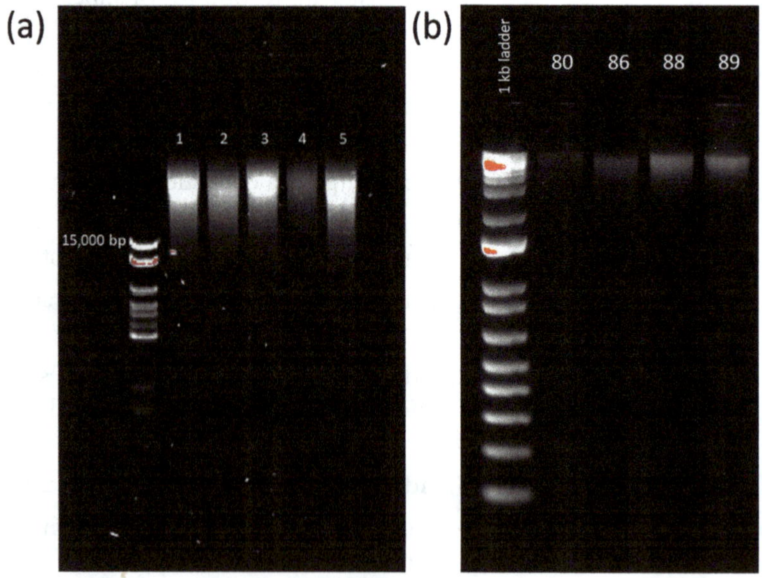

Fig. 1 Gel electrophoresis of extracted DNA from (**a**) wine ferment samples and (**b**) fermented palm sap samples. All genomic DNA work for the following library construction process

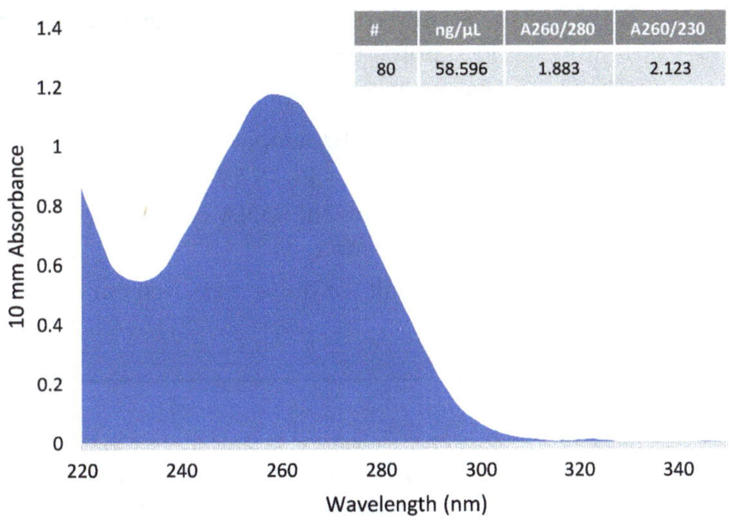

Fig. 2 DNA quality assessed by Nanodrop spectroscopy in sample number 80. Value A260/280 indicates the absorbance ratio of DNA purity that generally takes value at ~1.8–2.0 while A260/230 indicates less contaminants in the range of 2.0–2.2 (*see* **Note 7**)

Table 1
DNA quality and quantity from measurement by Nanodrop and Qubit (*see* Note 9)

	Nanodrop			
#Sample	**Concentration (ng/µL)**	**A260/280**	**A260/230**	**Qubit (ng/µL)**
80	58.596	1.883	2.123	33.6
86	34.965	1.908	2.273	17.5
88	29.284	1.929	2.262	15.1
89	52.801	1.921	2.195	29

3.4 DNA Quantity and Quality Measurement

3.4.1 Nanodrop

- Prepare sample for measurement in Nanodrop.
- Prepare a blank solution for Nanodrop measurement, using the elution buffer (*see* **Note 6**).
- Load 1 drop (approximately 1.5 µL) of blank, measure, and then follow with each sample of interest.
- The quality of DNA is visualized in Fig. 2 and the quantification values are in Table 1.

3.4.2 Qubit

- Prepare samples for measurement in Qubit.
- Prepare 2 tubes for 2 qubit standards (Std 1 and Std 2).
- Add 10 µL of each standard to each tube (*see* **Note 8**).

- Add 190 μL of Working solution (maximum amount is supposed to 200 μL).
- Vortex briefly.
- As for the sample, add 198 μL of working solution to the tube and add 2 μL of sample. Vortex briefly to mix.
- Measure the DNA concentration noting the 2 μL dilution on qubit machine.
- Result for qubit measurement is presented in Table 1.

4 Notes

1. Store in a chiller or ice before application to minimize the contamination and stabilize the chemical compounds.
2. The repetition will give a maximum binding of PVPP to reduce phenolic content. Time of repetition depends on the type of sample and possible content of polyphenols. More repetitions will be needed if more polyphenols are suspected.
3. Homogenizing sample and buffer lysis is crucial in this step. Substances in the PowerBead Tube ensure that the contaminants are dissolved and prevent DNA degradation.
4. SYBR safe is sensitive to the light, store in a dark container.
5. Mixing loading dye and DNA is important to ensure DNA is evenly distributed throughout the loading dye, which contains tracking dyes and a density agent. The tracking dyes help visualize the DNA migration during gel electrophoresis, while the density agent helps the DNA sink into the well of the gel. Proper mixing also prevents air bubbles from forming in the loading mixture, which can interfere with the migration of DNA fragments during electrophoresis.
6. Blank solution for Nanodrop can be C6 solution or DNA-free water.
7. Qubit only shows DNA concentration. Otherwise, Nanodrop can describe DNA purity by the absorbance ratio at 260/280 nm and 260/230 nm. The purity of DNA is at of ~1.8 and ~2.0 is for RNA. A 260/280 ratio of ~1.8–2.0 is generally accepted as "pure" for genomic DNA. If the ratio is out of range, contaminants such as protein and phenol may present that can absorb wavelength at or near 280 nm. The second ratio also indicates purity, which is commonly expected at 2.0–2.2. If the ratio is lower than expected, contaminants may absorb wavelength at 230 nm.
8. Qubit standards are light sensitive and must be stored at 4 °C.

9. If the yield of DNA concentration is low, put the eluted DNA into vacuum concentrator machine or conducting classical method like ethanol precipitation will be helpful. Here are general steps for ethanol precipitation:
   (a) Add 2.5 volumes of ice-cold ethanol absolute (>99%) to the DNA sample.
   (b) Add 0.1 volume of 3 M sodium acetate (pH 5.2) to the sample and mix well by inverting the tube several times.
   (c) Place the sample in a −20 °C fridge for 1–2 h to allow the DNA to precipitate.
   (d) Centrifuge the sample at maximum speed (14,000 × *g*) for 10–15 min at 4 °C.
   (e) Carefully remove the supernatant without disturbing the DNA pellet as much as possible.
   (f) Wash the pellet with ice-cold 75% ethanol to remove any remaining salt.
   (g) Centrifuge again at maximum speed (14,000 × *g*). Discard supernatant carefully and air-dry or vacuum-dry the pellet.
   (h) Resuspend the DNA pellet in an appropriate volume of sterile TE buffer or DNA-free water.

   The amount of ethanol and sodium acetate added can vary depending on the amount and type of DNA sample. It is important to use ice-cold ethanol and to centrifuge the sample at 4 °C to ensure maximum DNA precipitation.

Acknowledgments

INS acknowledges the support of an Australia Awards, supported by the Australian Government. The views and opinions expressed in this article are those of the authors and do not represent the views of the Australian Government.

References

1. Hill JE, Seipp RP, Betts M, Hawkins L, Van Kessel AG, Crosby WL et al (2002) Extensive profiling of a complex microbial community by high-throughput sequencing. Appl Environ Microbiol 68(6):3055–3066
2. Rastogi G, Sani RK (2011) Molecular techniques to assess microbial community structure, function, and dynamics in the environment. In: Ahmad I, Ahmad F, Pichtel J (eds) Microbes and microbial technology: agricultural and environmental applications. Springer, New York, pp 29–57
3. Riesenfeld CS, Schloss PD, Handelsman J (2004) Metagenomics: genomic analysis of microbial communities. Annu Rev Genet 38(1):525–552
4. Röling WF, Ferrer M, Golyshin PN (2010) Systems approaches to microbial communities and their functioning. Curr Opin Biotechnol 21(4):532–538

5. Hermans SM, Buckley HL, Lear G (2018) Optimal extraction methods for the simultaneous analysis of DNA from diverse organisms and sample types. Mol Ecol Resour 18(3): 557–569
6. de Lipthay JR, Enzinger C, Johnsen K, Aamand J, Sørensen SJ (2004) Impact of DNA extraction method on bacterial community composition measured by denaturing gradient gel electrophoresis. Soil Biol Biochem 36(10):1607–1614
7. Henderson G, Cox F, Kittelmann S, Miri VH, Zethof M, Noel SJ et al (2013) Effect of DNA extraction methods and sampling techniques on the apparent structure of cow and sheep rumen microbial communities. PLoS One 8(9):e74787
8. Pinto AD, Forte V, Guastadisegni MC, Martino C, Schena FP, Tantillo G (2007) A comparison of DNA extraction methods for food analysis. Food Control 18(1):76–80
9. Friar EA (2005) Isolation of DNA from plants with large amounts of secondary metabolites. In: Methods in enzymology. Academic Press, p 1–12
10. Kim CS, Lee CH, Shin JS, Chung YS, Hyung NI (1997) A simple and rapid method for isolation of high quality genomic DNA from fruit trees and conifers using PVP. Nucleic Acids Res 25(5):1085–1086
11. Fang G, Hammar S, Grumet R (1992) A quick and inexpensive method for removing polysaccharides from plant genomic DNA. BioTechniques 13(1):52–56
12. Bokulich NA, Mills DA (2012) Next-generation approaches to the microbial ecology of food fermentations. BMB Rep 45(7): 377–389
13. Aboul-Maaty NAF, Oraby HAS (2019) Extraction of high-quality genomic DNA from different plant orders applying a modified CTAB-based method. Bull Natl Res Cent 43(1):25
14. Paterson AH, Brubaker CL, Wendel JF (1993) A rapid method for extraction of cotton (*Gossypium* spp.) genomic DNA suitable for RFLP or PCR analysis. Plant Mol Biol Report 11(2): 122–127
15. Sahu SK, Thangaraj M, Kathiresan K (2012) DNA extraction protocol for plants with high levels of secondary metabolites and polysaccharides without using liquid nitrogen and phenol. ISRN Mol Biol 2012:1–6
16. Leiper KA, Stewart GG, McKeown IP, Nock T, Thompson MJ (2005) Optimising beer stabilisation by the selective removal of Tannoids and sensitive proteins. J Inst Brew 111(2):118–127
17. Durán-Lara EF, López-Cortés XA, Castro RI, Avila-Salas F, González-Nilo FD, Laurie VF et al (2015) Experimental and theoretical binding affinity between polyvinylpolypyrrolidone and selected phenolic compounds from food matrices. Food Chem 168:464–470
18. Gobert A, Evers MS, Morge C, Sparrow C, Delafont V (2022) Comparison of DNA purification methods for high-throughput sequencing of fungal communities from wine fermentation. MicrobiologyOpen 11(5):e1321
19. Liu D, Chen Q, Zhang P, Chen D, Howell KS (2020) The fungal microbiome is an important component of vineyard ecosystems and correlates with regional distinctiveness of wine. mSphere 5(4):e00534–e00520
20. Liu D, Howell K (2021) Community succession of the grapevine fungal microbiome in the annual growth cycle. Environ Microbiol 23(4): 1842–1857

Chapter 3

Fluorescence Spectroscopy for Red Wine Authentication

Ranaweera K. R. Ranaweera, Adam M. Gilmore, and David W. Jeffery

Abstract

The molecular fingerprint obtained from fluorescence spectroscopy has been used in combination with chemometrics for the authentication of red wine according to region, variety, and vintage. The approach relies on dilution of a small amount of wine prior to absorbance-transmission and fluorescence excitation-emission matrix (A-TEEM) analysis, which provides UV/Vis spectra and fluorescence landscapes for determining wine color parameters and for authentication. This protocol describes pre-operation checks, sample preparation and spectral analysis, data pre-treatment, and examination, along with parallel factor analysis for fluorophore assignment and extreme gradient boosting discriminant analysis for classification, while highlighting key points in the overall methodology.

Key words A-TEEM, Authenticity, Excitation-emission matrix, Extreme gradient boosting, Fluorophore, Machine learning, Parallel factor analysis, Provenance

1 Introduction

There are many approaches that can help detect wine fraud arising from adulteration or substitution of variety, region, or vintage [1]. Non-targeted spectroscopic techniques that provide a product "fingerprint" are attractive due to their rapid and user-friendly nature, with fluorescence spectroscopy being particularly appealing due to the high level of sensitivity and selectivity it offers [2]. Similar in some respects to UV/Vis spectroscopy, which can also be used for wine authentication, the fluorescence technique involves electronic transitions of molecules via absorption of light with wavelengths in the UV and visible range (190–800 nm). Being a type of luminescence, it relies on excitation at certain wavelengths (λ_{ex}), whereby molecules absorb photons, and detection of light emission at longer wavelengths (λ_{em}) from the thermally equilibrated excited state, with the release of photons rapidly returning the molecule to the ground electronic state [3].

Fluorescence spectroscopy detects fluorophores—in contrast to chromophores with UV/Vis—although these can typically be

María Ángeles Pozo-Bayón and Carolina Muñoz González (eds.), *Wine Analysis and Testing Techniques*,
Methods and Protocols in Food Science, https://doi.org/10.1007/978-1-0716-3650-3_3,

the same types of molecules, such as aromatic compounds and those with conjugated double bonds (phenolics, amino acids, vitamins, porphyrins [4]) typically present in grape and wine [5, 6]. The technique yields excitation and emission spectra that characterize the unique properties of fluorophores, either as individual spectra or as a fluorescence landscape in the form of an excitation-emission matrix (EEM) [4]. It is possible to collect NIST-traceable fluorescence EEMs by accurately correcting spectral properties and secondary factors associated with Rayleigh and Raman scattering, and inner filter effects [7]. Intact food systems (including opaque solids or turbid liquids) can be measured using the front-face technique (typically with 30° illumination angle), which analyses fluorophores in their native environment but can be prone to signal distortion by strong absorbance and light scattering. In contrast, the classical right-angle technique (0° illumination angle) is more sensitive and selective, especially when used with dilute samples where both absorbance and fluorescence intensities are proportional to the fluorophore concentration, that is, under Beer-Lambert linear conditions [3, 5].

Combining UV/Vis and fluorescence, a spectroscopic technique known as absorbance-transmission and fluorescence excitation-emission matrix (A-TEEM) uses the classical sample geometry to provide absorbance data along with fluorescence EEMs using an Aqualog instrument [8]. This provides five parameters that are compound (and solvent) specific: absorbance gives extinction and spectral shape information; fluorescence yields information on quantum efficiencies (relating absorbance to fluorescence intensity) along with excitation and emission spectra. Data from A-TEEM have been combined with chemometrics and machine learning algorithms to generate accurate classification models for wine authentication by grape variety, production region, and year of vintage [9–12]. This simple approach involves sample dilution, recording of spectral data with an Aqualog spectrophotometer, spectral correction, and unfolding of EEMs into a two-way array, data pre-processing and compression, then modeling based on sample class. The fundamental protocols are described for sample preparation, instrument operation for A-TEEM data acquisition, and data analysis including machine learning techniques associated with classification of red wine.

2 Materials

- Absolute ethanol, (UV Optical or HPLC chromatography grade), 37% analytical grade hydrochloric acid (HCl), and water purified through a Milli-Q purification system.

- Type 1FL (4-way clear, spectral range 200–2500 nm, path length 10 × 10 mm, chamber volume 3500 μL) macro fluorescence cuvette (Helma or Firefly Scientific).
- Glass storage vials with PTFE lined screw cap, 12 mL.
- External water bath or thermoelectrically cooled cuvette holder (optional; to equilibrate temperature and circulator for sample chamber to 20 °C, HORIBA Instruments Inc., Piscataway, NJ, USA).
- Aqualog UV-800C spectrofluorometer (HORIBA Instruments Inc.).
- Fast-01 (10 mL vials, up to 24) or Fast-02 (2 mL vials, up to 96) autosampling unit (HORIBA Instruments Inc.), each with an 80 μL flow cell with 1 cm transmission and 0.5 cm fluorescence optical path to facilitate inner filter effect (IFE) correction.

3 Methods

3.1 Acidified Ethanol Solution Preparation for Sample Dilution

1. Measure 500 mL of ethanol (UV Optical or HPLC grade) and 500 mL of ultrapure water (Milli-Q, 18.2 MΩ·cm) in a measuring cylinder and transfer to a 1 L beaker with a magnetic stirrer bar.
2. Add a few drops of 1.0 M HCl to adjust the pH to 2.0 using a pH electrode while stirring with a magnetic stirrer.
3. Use vacuum filtration apparatus and a 0.45 μM PTFE filter membrane or a 0.2–0.45 μM PTFE syringe-mounted filter. In either case, rinse the filter set-up with 100 mL of the acidified ethanol solution and discard it to ensure no extractable UV absorbing or fluorescing materials are present in filtrate (which could lead to a blank with unacceptable background signals). Filter the remaining solution and transfer the contents to a clean screw cap bottle.

3.2 Wine Sample Preparation

1. Transfer 1 mL of the wine to a 1.5 mL Eppendorf tube and centrifuge at 9300 rcf for 10 min. Dilute the sample by pipetting the required supernatant volume of the centrifuged sample (example dilutions indicated below) into a 12 mL glass vial and adding filtered 50% (v/v) ethanol pH 2.0 solution, to obtain a diluted sample that fits within the Beer-Lambert concentration relationship (*see* **Note 1**).

   Red wine: 150-fold dilution (40 μL wine: 5960 μL solvent).

   White wine: 50-fold dilution (120 μL wine: 5880 μL solvent)

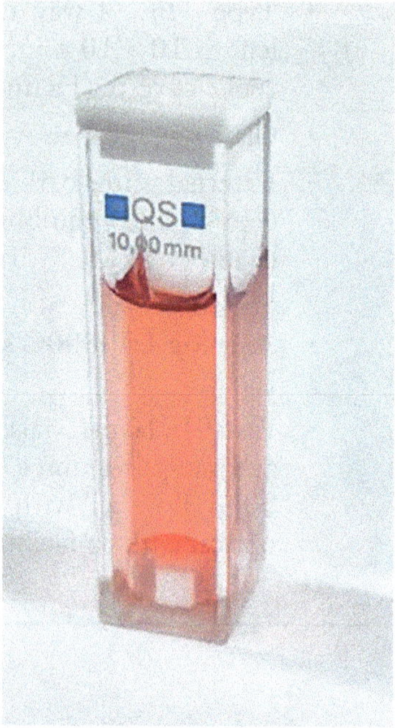

Fig. 1 4-Way clear 1-cm path length quartz fluorescence cuvette used for analysis of diluted red wine samples

2. Vortex the diluted and capped samples for 60 s and sonicate for 10 min to degas the mixture. Transfer 3000 μL of the degassed sample to a quartz fluorescence cuvette (Fig. 1) containing a mini magnetic stirrer bar.
3. Prepare a blank sample in the same manner by pipetting 3000 μL of 50% (v/v) ethanol pH 2.0 solution into a clean glass vial, vortex the capped vial for 60 s, sonicate for 10 min, and transfer 3000 μL to a quartz fluorescence cuvette containing a stirrer bar.

Samples can be prepared ahead of time to equilibrate and stored in acid-washed brown glass vials sealed with a screw cap. The preparation to analysis time lag should be limited (e.g., no more than 2 h) and consistent within a batch of samples for analysis to minimize the effects of oxidation.

3.3 Pre-operation Routine Calibration Check

1. To assess the wavelength calibration and instrument sensitivity, "Water Raman SNR and Emission Calibration" needs to be undertaken before starting the analysis.
2. Ensure the Aqualog has been ON for a minimum of 45 min and up to an hour to allow the lamp to warm up and stabilize.
3. Place the sealed water standard shown in Fig. 2a (10 mm quartz cuvette filled with high-pure water reference material, Starna Scientific Ltd.) into the sample chamber. It will click into place and be the same as shown in Fig. 2b.

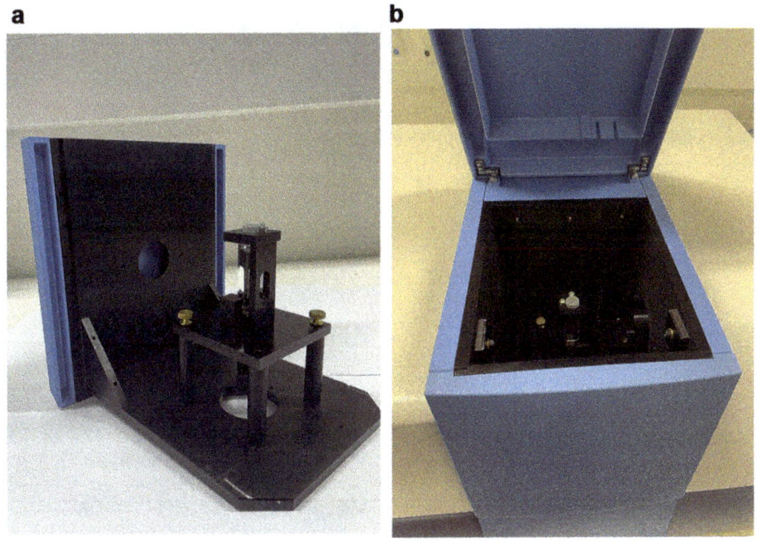

Fig. 2 Sealed water standard sample holder (**a**) side view and (**b**) in position within the instrument

Check cuvette for dust, fingerprints, etc., and wipe with lint free cloth and 96% ethanol if needed.

4. With the sealed water standard in position, run *Water Raman SNR and Emission Calibration* using the Aqualog (HORIBA Instruments Inc.) instrument control software (Collect > Aqualog Service Only). After around 60 s, the system will prepare a report assessing the peak position and the signal-to-noise ratio (SNR). The water Raman peak should occur at 397 nm and the SNR ratio must be above 20,000. A PASS/FAIL result will be provided for both parameters, as shown in Fig. 3. Monochromator adjustment may be required if the Raman peak wavelength fails and the xenon lamp may need changing if SNR fails. However, the quality of the water used for the calibration check is an important consideration and should be free of dissolved organic matter—general laboratory water is not sufficient.
5. Before analyzing samples, generate a Raman Scattering Unit (RSU) factor (RU icon in the Aqualog software) for fluorescence data normalization according to the water Raman peak area, using the sealed water standard and following settings:

   Integration time: 10 s.

   CCD Increments: 8 pixels (approx. 4.66 nm).

   Gain: Medium.
6. When the RSU analysis is complete, the normalization factor needs to be adjusted in the *RSU Adjust* tab of the workbook to match with the integration time to be used for Sample Q

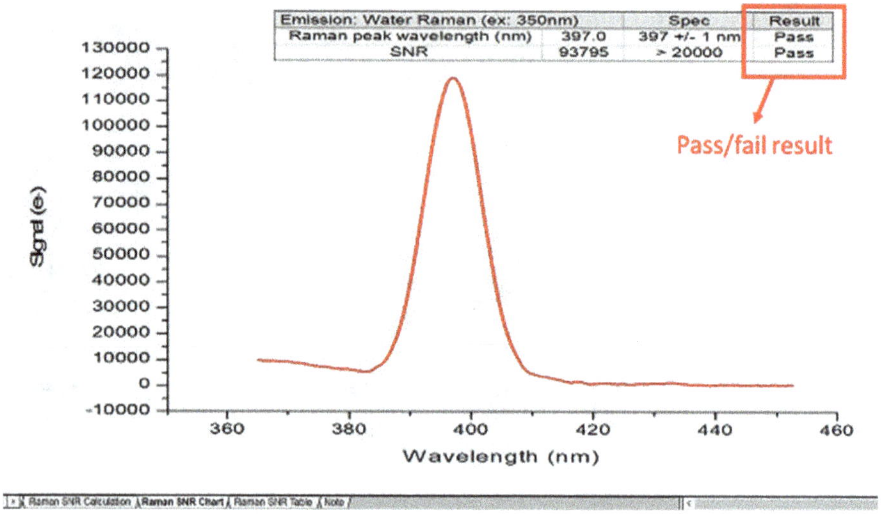

Fig. 3 Water Raman peak wavelength and signal-to-noise ratio (SNR) test results

operation (generally 0.1–0.2 s for red wine to avoid CCD detector saturation). The subsequent RSU Adjust area value will automatically recalculate for the new integration time, providing a normalization factor relevant to the sample set for post-processing the spectral data.

3.4 Sample Analysis for a Single Cuvette

1. The instrument sample compartment, sample and solvent materials should all be at room temperature (25 °C). After the pre-operational procedures, remove the water standard cuvette from the chamber and place the sample holder in the chamber as shown in Fig. 4. Test blank and sample cuvettes for optical cleanliness and absorbance and fluorescence baseline performance (*see* **Note 2**).
2. Prior to placing the cuvette in the sample holder chamber, check for air bubbles in the sample. If any bubbles are evident, carefully tap the cuvette to remove them before inserting the cuvette into the chamber shown in Fig. 4. Before data acquisition, stir the sample cuvette in the sample chamber for 1–2 min, and keep the stirrer on at medium speed (5–6) while running the analysis to ensure consistency within the sample. The speed at which the sample should be mixed depends on the viscosity of the sample (*see* **Note 3**).
3. Make sure that the sample chamber door is properly closed during the analysis to avoid any outside light entering the chamber.
4. Samples can be analyzed as a batch using "Sample Q method." Scan settings from the menu need to change as follows:

   Integration time: Integration (s) = 0.2 (that is the same integration time used to determine the RSU Adjust area).

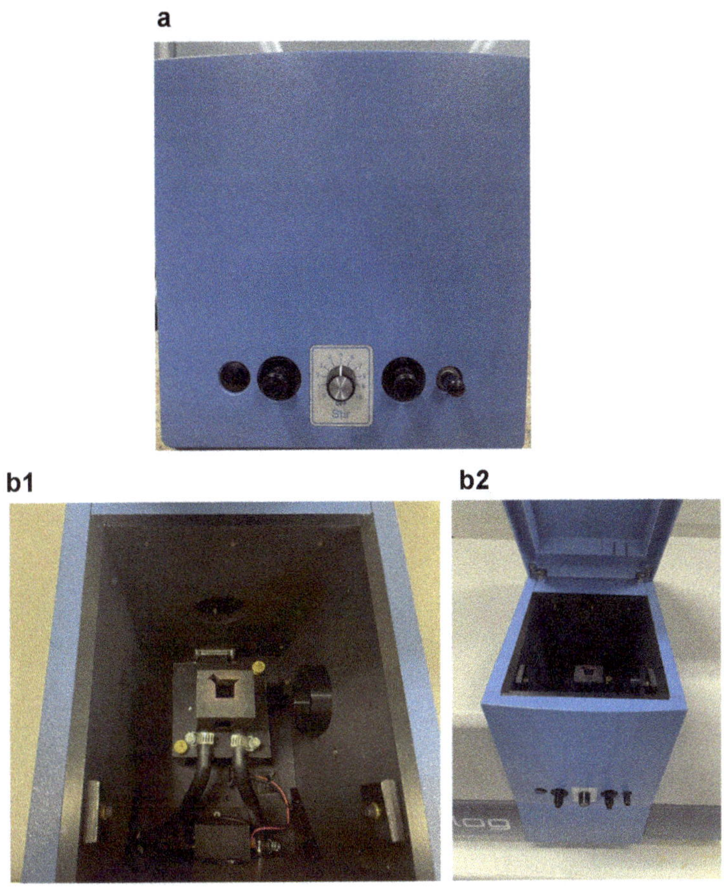

Fig. 4 Images of sample compartment showing (**a**) front view and (**b**) top view with location for the cuvette in the center within the instrument

Wavelength Settings:

Excitation Wavelength = 700 nm (High) and 240 nm (Low) with 5.00 nm (Increment).

Emission Wavelength = 829 nm (High) and 247 nm (Low) with 4.66 nm (8 pixels) (Increment).

Saturation mask width = 10 nm.

CCD Gain = Medium.

5. The Aqualog software allows for selection of different post-processing options (Data Processing menu) for IFE, Rayleigh masking (RM) first and second order (sum of slit widths should be 12 nm) and normalization (to the specific RSU Adjust area obtained earlier). There are EEM interpolation options to make the data comparable with other Aqualog instruments.
6. Sample analysis must initiate with a blank sample (*see* **Note 2**) and be followed with the diluted wine samples. Sample table configuration needs to be set up according to the number of

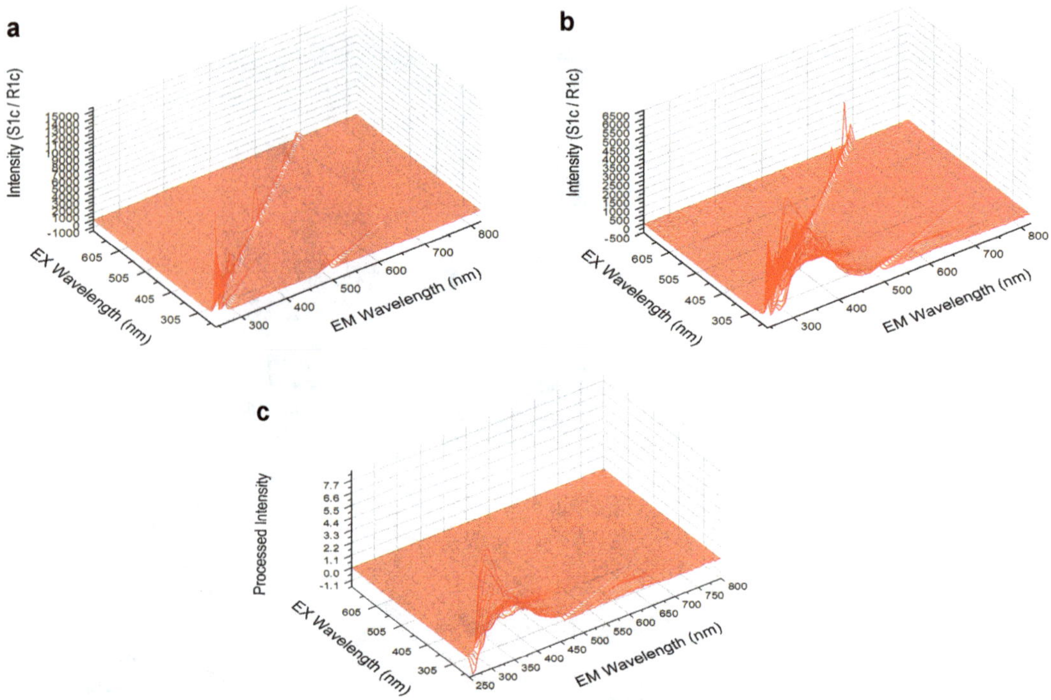

Fig. 5 Waterfall plot of (**a**) blank, (**b**) wine sample, and (**c**) wine sample corrected for IFE, RM, and normalized according to RSU

samples to be analyzed. Generally, for each 10 samples, a blank needs to be run prior to the analysis of another 10 samples. The Aqualog software will undertake the pre-processing of data at the time of data collection and provides the corrected EEM data based on IFE, RM, and normalization according to the post-processing options. Blanks, samples, and corrected EEMs can be visualized with waterfall plots (Fig. 5) to provide a convenient initial check of the data.

7. It is important to pay attention to indications of contamination (*see* **Notes 2** and **4**). Cuvettes and water or solvent/acid sources can have contaminations associated with them. Most sample contaminations are in the region of the excitation below 300 nm and emission below 400 nm (Fig. 6).

8. If there is any indication of contamination in the waterfall or contour plots, cuvette cleanliness or the purity of the dilution solvent needs to be checked, because any signal in the "blank" file will be automatically subtracted from the samples in the batch. Contamination can also be indicative if sample measurements have negative absorbance values. In that case, it is recommended to prepare the blank with a clean or new glass vial and rerun the batch. Ultimately, it is advisable to carefully check the blank sample (*see* **Notes 2** and **4**) to identify and rectify any issues prior to diluting wine samples for analysis.

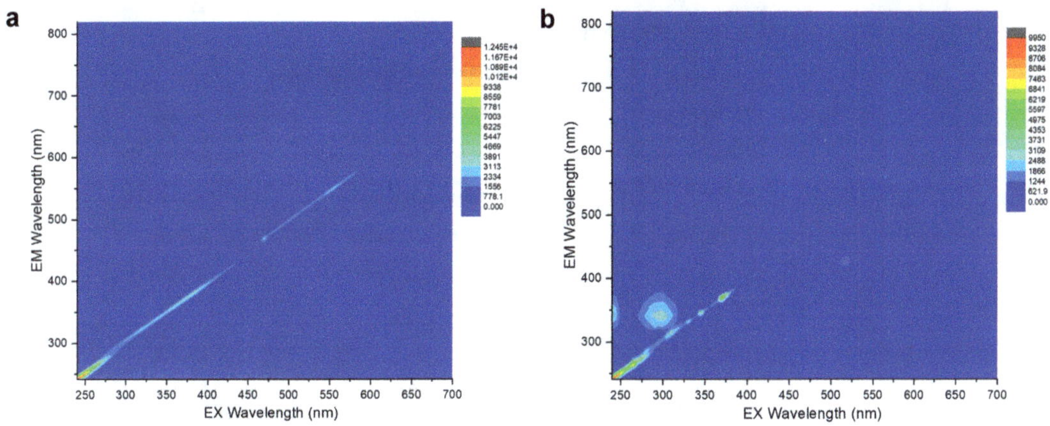

Fig. 6 Contour plots of EEM spectra for a blank run showing (**a**) uncontaminated blank and (**b**) contaminated blank

3.5 Sample Analysis for Multiple Samples with Fast-01 Autosampling Unit

1. Turn on unit, check connections to flow cell, waste, and rinse reservoir. Collect RSU unit as above (Subheading 3.3) using standard single-cell sample holder. Replace the sealed water standard sample holder with Fast-01 water-jacketed sample drawer and flow cell. In the Sample Q method, scan settings need to enable for the "Injector." On the injector, set up the method as in Fig. 7.
2. Transfer 10 mL aliquots of blank solvent and diluted samples to new clean vials, seal with cap/septum and place in sample racks for analysis (*see* **Note 4**). Fit the cooling lid cover in place and close lid to maintain temperature and darkness for sample vials. Review details in the configuration, vial trays, and blank/sample groups as required for the batch analysis.
3. Before running the samples, fill the rinse reservoir with 50% ethanol pH 2 solvent and Run:
   (a) Wash Needle (1×).
   (b) Wash Loop (3×).
4. When analysis is complete, replace and fill the rinse reservoir with suitable solvent (e.g., 20% v/v aqueous isopropanol) and run as above to clean the autosampler.

3.6 Data Processing

Origin OEM software (OriginLab Corporation, Northampton, MA, USA) incorporated into the Aqualog software undertakes processing based on the method settings and provides corrected data based on IFE, RM, and normalization. A general check of absorbance and fluorescence data integrity and evaluation of outliers should be undertaken prior to further treatment of the data (*see* **Note 5**).

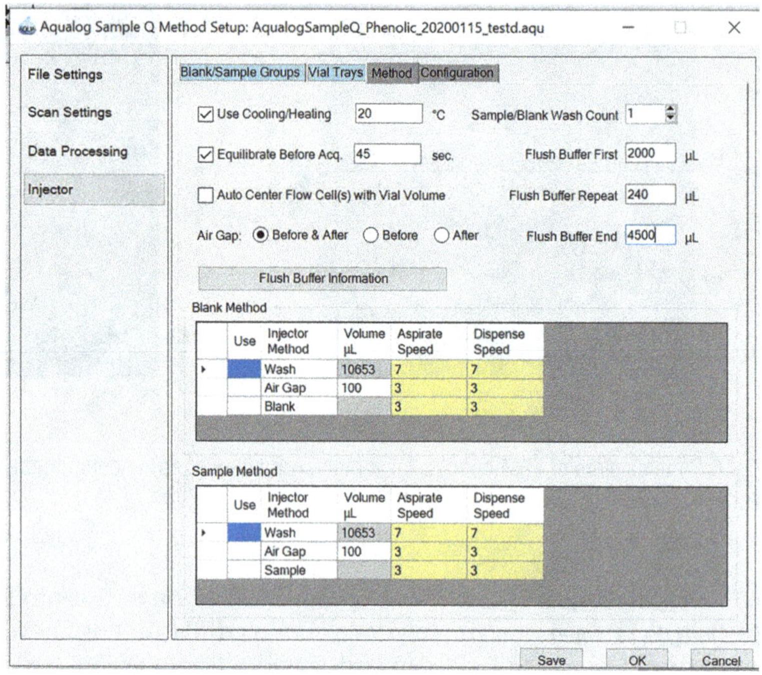

Fig. 7 Aqualog Sample Q method set up for multiple sample analysis using the Fast-01 autosampler

3.6.1 PARAFAC Modeling

EEM data obtained from the Aqualog is arranged in three-way arrays; sample × excitation wavelength × emission wavelength (3D). Using Solo software package (Eigenvector Research, Inc., Manson, WA, USA), parallel factor analysis (PARAFAC) is used to decompose EEMs and facilitate the evaluation of putative fluorophore identities according to the components in the sample. First-order and second-order Rayleigh filters are applied to EEM data (e.g., ± 16 nm and ± 32 nm, respectively) based on sample properties. The filtered values (greater than or equal to the values set in the Aqualog software for RM, Subheading 3.4) can be replaced by the software with either missing data or interpolation from either side of the scatter band. EEM data is normalized to an area of one and non-negativity constraints are applied to intensity, emission, and excitation wavelengths (nm). The number of components required for the model can be selected; typically, wine data can be analyzed using four or five components. To evaluate and validate PARAFAC models, it is recommended to follow tutorial publications [13, 14]. Conventionally, the key goodness of fit parameters for PARAFAC models include the core consistency values and split-half matching analysis. Core consistency close to 100% is optimal for a stable model or a similarity above 90% in the split half analysis indicates the same (Fig. 8).

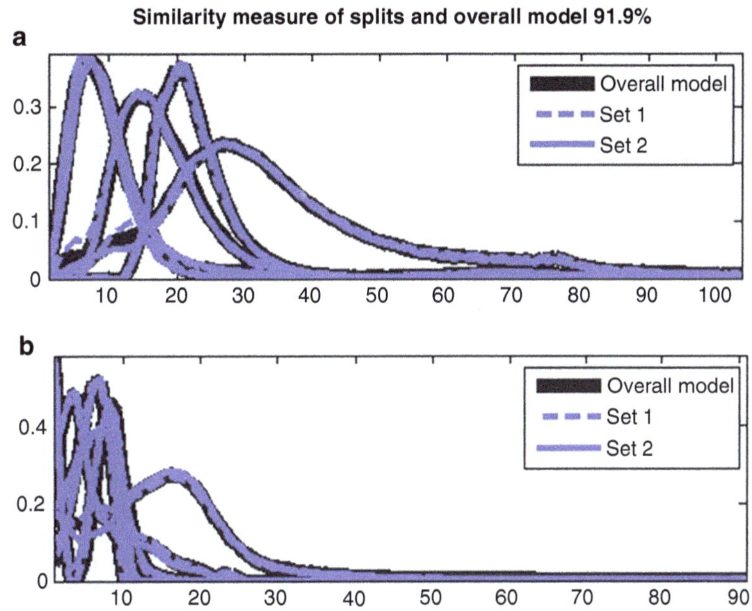

Fig. 8 Split-half analysis for (**a**) excitation and (**b**) emission of the five components from the PARAFAC model obtained using EEM data, showing the overall model plus two subsets of the data (Set 1 and Set 2). The x-axis shows the wavelength channel number (based on the excitation and emission ranges used for the analysis) and the y-axis shows the loading values. The overlap of the overall model and Sets 1 and 2 shows that very similar excitation and emission loadings are obtained in each case, thus verifying the model is stable with the chosen number of components. (Reprinted from Ranaweera et al. [10]. Copyright (2021), with permission from Elsevier)

3.6.2 Classification Modeling

1. Prior to classification modeling of wine samples according to grape variety, production region, and year of vintage, 3D EEM data need to be reshaped to 2D (excitation wavelength values multiplied by emission wavelength values) (*see* **Note 6**). For unsupervised pattern recognition, cluster analysis with partitional k-means can be applied and exploratory data analysis can be carried out with principal component analysis (e.g., [11]).
2. For classification, supervised multivariate methods can be employed for wine authentication purposes, such as partial least squares discriminant analysis (PLS-DA), linear discriminant analysis (LDA), and support vector machine (SVM). Classification models built with the extreme gradient boosting (XGBoost) algorithm (Fig. 9), based on a boosted decision tree approach, tend to yield higher accuracy in classification (e.g., [9]), although the potential for overfitting with supervised machine learning techniques needs to be controlled [15, 16].

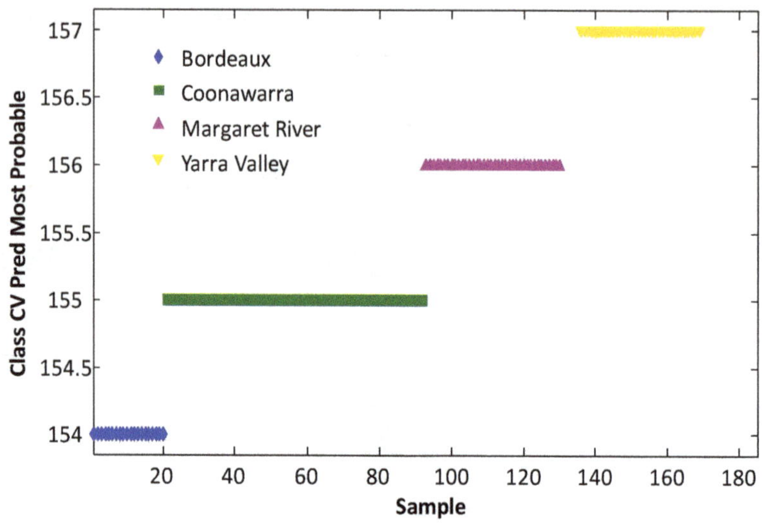

Fig. 9 XGBDA analysis of EEM data for Cabernet Sauvignon wines from different regions showing error-free classifications in class CV predicted most probable for Bordeaux, Coonawarra, Margaret River, and Yarra Valley. (Reprinted from Ranaweera et al. [9]. Copyright (2020), with permission from Elsevier)

3. Using Solo software (see https://wiki.eigenvector.com/index.php?title=Main_Page for user guides), application of XGBoost discriminant analysis (XGBDA) for unfolded EEM data first involves PLS compression using 10 latent variables, with mean centering pre-processing and decluttering with generalized least squares weighting at 0.2 to both calibrate and cross-validate (k = 10, Venetian blinds procedure). With large enough data sets, data can be split into training and test sets of an appropriate proportion, such as based on the number of features in the model (e.g., [17]) to assist with model optimization and validation. ASTM E2617 also offers recommendations for minimum sample numbers for calibration and validation sets [18]. The XGBDA modeling occurs with the xgboost algorithm and gbtree booster with basic parameters applied as follows: eta = 0.1, max_depth = 2, and num_round = 300.
4. To maximize performance of the model, absorbance and EEM data from A-TEEM can be combined into a multi-block data set (e.g., using Solo + MIA software). The combination (incorporating the five parameters mentioned in the Introduction) has shown a slight improvement in score probabilities for regional classification [12] and was useful for precise sub-regional classification [11], due to the better supporting the decomposition of specific components present in the samples (in other words, enhancing the molecular fingerprint). A grid search can be undertaken to tune the pre-processing

Table 1
Confusion matrix results for XGBDA analysis of multi-block data (absorbance + EEM) for Australian Cabernet Sauvignon wines according to their GI[a]

Region	N	Sensitivity%	Specificity%	Error%	Precision%	F1 Score
Barossa Valley	56	100.00	100.00	0.00	100.00	1.00
Clare Valley	16	100.00	100.00	0.00	100.00	1.00
Eden Valley	24	100.00	100.00	0.00	100.00	1.00
Frankland River	14	100.00	100.00	0.00	100.00	1.00
Langhorne Creek	50	100.00	100.00	0.00	100.00	1.00
Margaret River	34	100.00	100.00	0.00	100.00	1.00
McLaren Vale	50	98.00	100.00	0.23	100.00	0.99
Riverland	168	100.00	99.64	0.23	99.41	0.99
Wrattonbully	20	100.00	100.00	0.00	100.00	1.00
Murray Darling	10	100.00	100.00	0.00	100.00	1.00

Data from Ranaweera et al. [10]. Copyright (2021), with permission from Elsevier

[a]Sensitivity (true positive rate): proportion of positive cases that were correctly identified = 100 × TP/(TP + FN); Specificity (true negative rate): proportion of negative cases that were classified correctly = 100 × TN/(TN + FP); Misclassification error: proportion of samples which were incorrectly classified = 100 × (1-accuracy), where accuracy = (TP + TN)/(TP + TN + FP + FN); Precision: proportion of positive cases giving a true positive result = 100 × TP/(TP + FP); F1 Score: harmonic mean of precision and sensitivity = 2TP/(2TP + FP + FN)

options and hyperparameters during model optimization, either using the Solo model optimization tool (see https://wiki.eigenvector.com/index.php?title=Modeloptimizergui) or undertaking this manually.

5. Class predictions from the XGBDA model can be selected based on classification rules "class predict strict" (probability > threshold value of 0.5) or "class predict most probable" (highest probability). Confusion matrices are used to evaluate the performance of the models based on values for true positive (TP), false positive (FP), true negative (TN), and false negative (FN) for each class. For example, Table 1 shows the confusion matrix results for classifying Cabernet Sauvignon wines according to their geographical indication (GI) using fused EEM and absorbance data sets.

4 Notes

1. A primary condition for accurate A-TEEM spectroscopy of wine (or grape extracts) is to ensure all dissolved components are in the linear Beer-Lambert absorbance range from around 240–530 nm. This can be accomplished by a making a serial

dilution and plotting the absorbance values at all wavelengths as a function of dilution factor. A general rule of thumb for most chromophoric dissolved natural substances in wine is that A280 must be at least 0.2 but not more than 1.0 OD cm^{-1}. For most intense red wines, this typically involves dilution of at least 100-fold, and for white wines at least 50-fold.

2. When using cuvettes (4 mL, 4-way clear quartz, transmission 200–2500 nm) for analysis it is recommended to keep at least one "check" cuvette free from sample exposure (only ever containing dilution solvent but well cleaned and carefully stored between uses) for performing the sample cuvette cleanliness check. Conduct a standard blank A-TEEM scan on the check cuvette then use a previously cleaned cuvette, that is one to be used for further sample evaluation but filled only with solvent for the "sample" scan. Evaluate the resulting absorbance spectrum: the absorbance mean should be 0 cm^{-1} with a baseline noise envelop of <0.01 AU and no significant systematic deviations from the mean. Additionally, inspect the processed EEM contour plot to ensure all regions of the EEM are free of any significant positive or negative deviations, that is only exhibiting random noise with a mean of 0. When possible, it is also recommended to use the same dilution solvent and cuvette for the blank measurement, then spike and mix in the wine (or grape extract) into that solvent with stirring for the sample measurement; this helps subtract and eliminate cuvette- and solvent-dependent artifacts.
3. It is recommended to start the stirrer motor on a slow speed, watching to ensure the magnetic stirrer bar engages and is stirring correctly. If the stirrer bar begins "jumping" this can disturb the signal being recorded.
4. Similar to working with cuvettes as explained in **Note 2**, when using an autosampler unit (Fast 01 or Fast 02) it is advisable prior to the first runs of samples to thoroughly prime and rinse the system and flow cell with dilution solvent and prepare a blank solvent vial in duplicate. Measure the blank with the first vial and use the second vial for the "sample" to evaluate the signal baseline for absorbance and fluorescence, as explained in **Note 2**. Repeat this process with care to purge all air bubbles if baseline specifications are not met. It is also critical that the solvent used for rinsing between samples matches the solvent used to dilute/dissolve the samples; ideally, the solvent batches would be prepared together. Mismatching solvents can lead to refractive index changes at the sample-solvent interfaces as well as possibly thermal reactions, which can lead to poor and variable sample elution properties.

5. After collecting a set of A-TEEM scans it is highly recommended to separately import the time-date stamped Processed EEM (*PEM.dat) 3-way files and absorbance (*ABS.dat) files into Eigenvector Solo as X-block data sets to inspect for baseline issues and other possible outlier conditions. The baseline integrity of the absorbance spectra files is very important because these values are incorporated into the inner filter effect correction algorithm for the PEM files. It is possible to remove individual outliers by excluding them within the Edit Data window, but care should be taken to identify, if possible, the reason for outlier behavior. The absorbance spectra should all show a similar baseline with a value around 0 OD cm^{-1} at wavelengths >700 nm. Any significant deviations of the baseline or offset for the absorbance spectra could be a reason to reject both the *ABS.dat and *PEM.dat data sets and diagnose and remedy the cause of the problem. The most common causes of baseline issues are: (1). Air bubbles in the cuvettes or Fast-01 flow cell; (2). Mismatching of blank and sample solvents; (3). Improper placement of the cuvettes or flow cell in the optical path; (4). Light leaks into the sample compartment; (5). Mechanical issues with the instrument or autosampler.
6. When unfolding the three-way *PEM.dat file set (X-block) into a 2-way array, it is important to use the "Unfold Multiway" command in the Edit Data window and select/enter one mode to remain unfolded. This preserves the excitation × emission wavelength coordinate metadata as well the data ranges masked or excluded in the 3-way data format. It is also important to check that any classification ID information assigned in the 3-way data set is properly preserved and or re-entered after the unfolding step, before saving the 2-way X-block. If the class information is altered or deleted during unfolding, the Delete Class option can be used to re-enter the correct values.

Acknowledgments

We are immensely grateful to Associate Professor Susan Bastian and Dr. Dimitra Capone from the University of Adelaide for their contributions to the original research papers. R.K.R.R. is appreciative of the financial support provided by an Australian Government Scholarship and Wine Australia supplementary scholarship (WA Ph1909).

References

1. Ranaweera RKR, Souza Gonzaga L, Capone DL, Bastian SEP, Jeffery DW (2021) Authenticity and traceability in the wine industry: from analytical chemistry to consumer perceptions. In: Cifuentes A (ed) Comprehensive foodomics, 1st edn. Elsevier, Amsterdam, pp 452–480. https://doi.org/10.1016/B978-0-08-100596-5.22876-X
2. Ranaweera RKR, Capone DL, Bastian SEP, Cozzolino D, Jeffery DW (2021) A review of wine authentication using spectroscopic approaches in combination with chemometrics. Molecules 26(14):4334
3. Lakowicz JR (2006) Principles of fluorescence spectroscopy, 3rd edn. Springer, New York
4. Christensen J, Nørgaard L, Bro R, Engelsen SB (2006) Multivariate autofluorescence of intact food systems. Chem Rev 106(6):1979–1994. https://doi.org/10.1021/cr050019q
5. Airado-Rodríguez D, Durán-Merás I, Galeano-Díaz T, Wold JP (2011) Front-face fluorescence spectroscopy: a new tool for control in the wine industry. J Food Compos Anal 24(2):257–264. https://doi.org/10.1016/j.jfca.2010.10.005
6. Armstrong CEJ, Gilmore AM, Boss PK, Pagay V, Jeffery DW (2023) Machine learning for classifying and predicting grape maturity indices using absorbance and fluorescence spectra. Food Chem 403:134321. https://doi.org/10.1016/j.foodchem.2022.134321
7. Gilmore AM (2014) How to collect National Institute of Standards and Technology (NIST) traceable fluorescence excitation and emission spectra. In: Engelborghs Y, Visser AJWG (eds) Fluorescence spectroscopy and microscopy: methods and protocols. Humana Press, Totowa, pp 3–27
8. Quatela A, Gilmore AM, Gall KES, Sandros M, Csatorday K, Siemiarczuk A et al (2018) A-TEEM™, a new molecular fingerprinting technique: simultaneous absorbance-transmission and fluorescence excitation-emission matrix method. Methods Appl Fluoresc 6(2):027002. https://doi.org/10.1088/2050-6120/aaa818
9. Ranaweera RKR, Gilmore AM, Capone DL, Bastian SEP, Jeffery DW (2021) Authentication of the geographical origin of Australian Cabernet Sauvignon wines using spectrofluorometric and multi-element analyses with multivariate statistical modelling. Food Chem 335:127592. https://doi.org/10.1016/j.foodchem.2020.127592
10. Ranaweera RKR, Gilmore AM, Capone DL, Bastian SEP, Jeffery DW (2021) Spectrofluorometric analysis combined with machine learning for geographical and varietal authentication, and prediction of phenolic compound concentrations in red wine. Food Chem 361:130149. https://doi.org/10.1016/j.foodchem.2021.130149
11. Ranaweera RKR, Bastian SEP, Gilmore AM, Capone DL, Jeffery DW (2023) Absorbance-transmission and fluorescence excitation-emission matrix (A-TEEM) with multi-block data analysis and machine learning for accurate intraregional classification of Barossa Shiraz wine. Food Control 144:109335. https://doi.org/10.1016/j.foodcont.2022.109335
12. Ranaweera RKR, Gilmore AM, Bastian SEP, Capone DL, Jeffery DW (2022) Spectrofluorometric analysis to trace the molecular fingerprint of wine during the winemaking process and recognise the blending percentage of different varietal wines. OENO One 56(1):189–196. https://doi.org/10.20870/oeno-one.2022.56.1.4904
13. Murphy KR, Stedmon CA, Graeber D, Bro R (2013) Fluorescence spectroscopy and multi-way techniques. PARAFAC Anal Methods 5(23):6557–6566. https://doi.org/10.1039/C3AY41160E
14. Stedmon CA, Bro R (2008) Characterizing dissolved organic matter fluorescence with parallel factor analysis: a tutorial. Limnol Oceanogr Methods 6(11):572–579. https://doi.org/10.4319/lom.2008.6.572
15. Vabalas A, Gowen E, Poliakoff E, Casson AJ (2019) Machine learning algorithm validation with a limited sample size. PLoS One 14(11):e0224365. https://doi.org/10.1371/journal.pone.0224365
16. Ying X (2019) An overview of overfitting and its solutions. J Phys Conf Ser 1168(2):022022. https://doi.org/10.1088/1742-6596/1168/2/022022
17. Joseph VR (2022) Optimal ratio for data splitting. Stat Anal Data Min: ASA Data Sci J 15(4):531–538. https://doi.org/10.1002/sam.11583
18. ASTM (2017) Standard practice for validation of empirically derived multivariate calibrations. In: Molecular spectroscopy and separation science; surface analysis. ASTM International

Chapter 4

Mass Spectrometry-Based Methods for the Characterization of Wine Flavonoids

Ignacio García-Estévez, Erika Salas, María Teresa Escribano-Bailón, and Cristina Alcalde-Eon

Abstract

Mass spectrometry is one of the most sensitive techniques that can be used in wine analysis. The use of different mass spectrometric approaches allows one to obtain a clearer vision of the chemical complexity of wine. There are different mass spectrometry approaches that can be used for flavonoid characterization, such as mass spectrometry using electrospray ionization (ESI), which can be useful for both, identify and quantify flavonoids, or matrix-assisted laser desorption ionization-time of flight (MALDI-TOF) mass spectrometry that allows characterizing complex mixtures of flavonoids.

Key words Anthocyanin, Flavonol, Flavanol, ESI-MS, MALDI-TOF, Quantification, Identification, Phenolic compounds, Wine analysis

1 Introduction

Flavonoids are the most important compounds regarding the organoleptic properties of wine, such as color, bitterness or astringency. Mass spectrometry (MS) equipped with mild ionization techniques, such as electrospray ionization (ESI-MS), can be used to perform the qualitative and quantitative study of flavonoids in wine. In ESI-MS, the ionization process is done at room temperature and at atmospheric pressure. ESI-MS also allows the detection of multicharged ions in the positive ion mode ($[M+zH]^{z+})/z$ or negative ion mode ($[M-zH]^{z-})/z$. Besides, increasing the voltage in the ion source favors the formation of fragment ions (MS-MS), which gives supplementary information about the structure of the molecule. Fragment ions can be a very useful tool for determining molecular structure as they arise from the well-determined breaking of chemical bonds. However, the structural complexity of flavonoids makes this a difficult task. For instance, with regards to wine flavan-3-ols (also known as flavanols), their structural features make

María Ángeles Pozo-Bayón and Carolina Muñoz González (eds.), *Wine Analysis and Testing Techniques*, Methods and Protocols in Food Science, https://doi.org/10.1007/978-1-0716-3650-3_4,

their characterization in wine samples a difficult task. These compounds are formed by four basic sub-units, which are isomeric two by two, that is, (epi)catechins ((E)C) and (epi)gallocatechins ((E)GC), that can polymerize through different types of bonds, giving rise the so-called proanthocyanidins. The most abundant proanthocyanidins result from the condensation of flavan-3-ols units by a B-type bond, which is C–C bound that can occur between the C4 of the upper unit and the C8 or C6 of the lower unit. Furthermore, proanthocyanidins can be divided into two different groups: procyanidins, which are formed just by (epi)-catechin units and prodelphinidins, which contain in their structure (epi)gallocatechin units, only or along with (epi)catechin units. Within this last group, we could differentiate between *pure* prodelphinidins, which contain just (epi)gallocatechin units, and mixed or double mixed oligomeric prodelphinidins, which contain one (epi)-gallocatechin or two (epi)gallocatechin units in their structure along with (epi)catechin units. Moreover, a galloyl residue can be linked to the hydroxyl group at C3 of the epi(gallo)catechin units. As a result, for a given polymerization degree, that is, the number of sub-units forming the proanthocyanidins, the number of different structures of flavan-3-ol that can exist in wine is high.

Along with this high chemical similarity among flavonoid compounds, the complexity of wine samples makes it advisable to combine MS analysis with a previous separation technique, usually high-performance liquid chromatography (HPLC), which, in turn, can be equipped with a diode-array detector to provide additional information (HPLC-DAD). In those cases, mass spectrometers are mostly coupled to the chromatography systems via the UV-vis cell outlet. Thus, with this combination of techniques the different wine constituents will be separated in the HPLC system depending on their affinity to the stationary or to the mobile phase and will enter the mass spectrometer quite isolated, allowing the analysis of a large number of wine constituents, including minority compounds that would probably have been ignored in a direct injection of the sample into the mass spectrometer. HPLC-DAD-ESI-MS has been widely used for flavonoid characterization since it provides the means to make peak assignments and acquire quantitative data by relying on mass spectral data.

This approach is very complementary to MALDI-TOF (Matrix-Assisted Laser Desorption Ionization– Time of Flight) analysis, which is the ideal technique for the analysis of complex mixtures as it produces only a singly charged molecular ion for each parent molecule. Thus, complex mixtures of polyphenols from fruits or beverages can be analyzed by MALDI-TOF, such as the native flavonoids from grape that are still present in wine, which can also change and react with other native polyphenols and be transformed (during wine-making and aging) in the complex mixture of derived flavonoids present in wine.

An important difference between ESI-MS and MALDI-TOF is that, sometimes, when analyzing highly concentrated flavonoid solutions (even pure standards) you can detect stacking in ESI-MS, which is double (or sometimes even triple) of the flavonoid *m/z*. This fact can be related to the high concentration of the analyte in the ion source. Stacking has not been reported to occur in MALDI-TOF analysis. Although MALDI-TOF is usually used in proteomics analysis, its application in wine analysis is not yet widely developed.

In this chapter, the main characteristics of different methodologies set up for flavonoid characterization in wine samples based on both ESI and MALDI-TOF mass spectrometry are described.

2 Materials

2.1 Simultaneous Analysis of Anthocyanins and Anthocyanin-Derived Pigments and Flavonols by HPLC-DAD-ESI-MS^n

- Water purified through a Milli-Q purification system and acidified water (pH adjusted to 1.4 using 37% hydrochloric acid—HCl, analytical grade), 0.1% aqueous trifluoracetic acid (TFA) and acetonitrile (HPLC-grade).
- Malvidin 3-*O*-glucoside (purity greater than 95%) and quercetin 3-*O*-glucoside (purity greater than 95%).
- High-performance liquid chromatography (HPLC) equipment equipped with a binary or quaternary pump, a column thermostat, and a Diode-Array-Detector (DAD) coupled to a mass spectrophotometer equipped with a triple-quadrupole and/or linear ion trap mass analyzer and with an ESI source (ESI-MS).
- Stationary phase: C-18 reversed-phase column (for instance: a 5 μm, 150 mm × 4.6 mm column).

2.2 Flavan-3-ol Characterization by HPLC-MRM-MS

- Cationic solid phase extraction (SPE) cartridge (for instance, Oasis MCX cartridges from Waters Corp., Milford, MA, USA).
- Methanol (for analysis or HPLC grade), 37% hydrochloric acid (HCl, analytical grade), water purified through a Milli-Q purification system, formic acid 98% (HPLC grade), and acetonitrile (HPLC grade).
- Chlorogenic acid (primary reference standard).
- HPLC system equipped with a binary or quaternary pump and a column thermostat, coupled to a mass spectrophotometer equipped with a triple-quadrupole mass analyzer and an ESI source (ESI-MS).
- C-18 column with superficially porous, core-shell particles (for instance: a 2.7 μm, 150 mm × 4.6 mm column).

2.3 Analysis of Flavonoids by MALDI-TOF

- Methanol (analytical grade), acetone (analytical grade), water purified through a Milli-Q purification system, and acetic acid (analytical grade).
- C18 cartridge (for instance, Sep Pak tC18 cartridge—environmental model from Waters, Milford, MA, USA).
- Trans-3-indolacrylic acid (tIAA) matrix (purity ≥98.5%).
- Cation exchange resin (for instance, Dowex 50X8-400 from Thermo Fisher, Waltham, MA, USA).
- MALDI-TOF mass spectrometer employed in reflectron ion detection mode (for instance, a Bruker UltrafleXtreme with laser SmartBeamIII from Bruker, Bremen, Germany).
- Target plate ground steel (for instance, MTP 394 BC from Bruker, Bremen, Germany).

3 Methods

3.1 Simultaneous Analysis of Anthocyanins and Anthocyanin-Derived Pigments and Flavonols by HPLC-DAD-ESI-MS^n

3.1.1 Sample Preparation

1. Dilute (*see* **Note 1**) wine samples with acidified water (pH next to 1.4) to shift the equilibrium of anthocyanin and anthocyanin-derived pigments towards flavylium forms (positively charged).
2. Filter the diluted samples with 0.45 μm filters before the injection into the chromatograph.

3.1.2 HPLC-DAD Analysis

1. Inject the filtered samples into the HPLC-DAD system coupled to ESI-MS working with previously optimized conditions (see for example Alcalde-Eon et al. [1]) (*see* **Note 2**). Given the structural and chromatographic behavioral similarity between wine anthocyanins and flavonols, the analysis of both types of flavonoids can be done in a single injection.
2. Use the information supplied by the chromatograms recorded at 520 nm for the qualitative and quantitative analysis of anthocyanins (*see* **Note 3**). Use a calibration curve of malvidin 3-*O*-glucoside previously injected in the same conditions to quantify them (*see* **Note 4**).
3. Employ the information supplied by the chromatograms recorded at 360 nm for the qualitative and quantitative analysis of flavonols. Use a calibration curve of quercetin 3-*O*-glucoside previously injected in the same conditions to quantify them (*see* **Note 4**).

3.1.3 MS Analysis

In the HPLC-DAD-ESI-MS^n analyses of anthocyanins and flavonols, MS is usually employed for qualitative purposes (for obtaining the *m/z* of the molecular or protonated ion and fragmentation pattern of the individual compound) and rarely for quantitative purposes due, above all, to the absence of adequate internal standards.

1. Optimize the conditions of the mass spectrometer with malvidin 3-*O*-glucoside (the main monomeric anthocyanin in wines made with *Vitis vinifera* L. red grapes and the main source of anthocyanin-derived pigments). Positive ion mode is recommended for the analysis of anthocyanins and anthocyanin-derived pigments as they are already positively charged in acidic solutions.
2. Check that the ionization conditions in the ESI source allow an adequate ionization of the targeted compounds without causing them an extensive fragmentation. In the conditions optimized for anthocyanins, flavonols can be also detected in the form of protonated ions in the full mass spectra (*see* **Note 5**).
3. Use the combined information from HPLC-DAD-MS analysis, which will be supplied for each peak in the chromatogram recorded at the selected wavelength and the UV-vis spectra and mass spectra, along with the elution order and retention time for identifying anthocyanins or flavonols.
4. It is highly recommended to carry out fragmentation of the compounds in MS^2 and MS^3 analyses to obtain more information about their structure since some of the anthocyanins, anthocyanin-derived pigments, and flavonols present in wine can show the same *m/z*. For a given compound and in a given condition, the compound will be fragmented following a certain pattern, which is called fragmentation pattern.

3.1.4 Fragmentation Patterns of Anthocyanins, Anthocyanin-Derived Pigments and Flavonols

1. To establish the fragmentation pattern, annotate the *m/z* of the molecular ion (anthocyanins) or protonated ion (flavonols) in the full mass spectra at the retention time of the targeted peak at 520 or 360 nm. Then, annotate the *m/z* of all of the fragment ions appearing in the MS^2 spectra and calculate the differences between them and the *m/z* of the molecular ion. These will be the *neutral losses* between the molecular ion and the fragment ions in MS^2. Annotate the *m/z* of the fragment ions in MS^3 and calculate the differences between them and the fragment ion of MS^2 from which they originate.
2. Localize the most abundant ion in MS^2 spectra to identify the anthocyanidin or the flavonol aglycone (Fig. 1) from which grape native anthocyanins or flavonols respectively derive. Additionally, the identity can be confirmed from the results of the MS^3 analysis where the aglycone is fragmented and

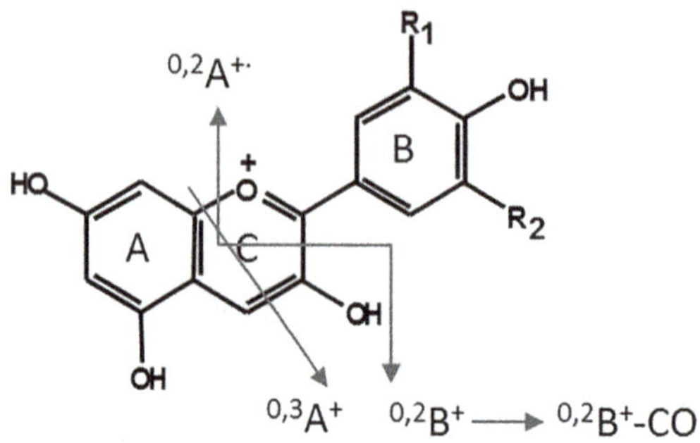

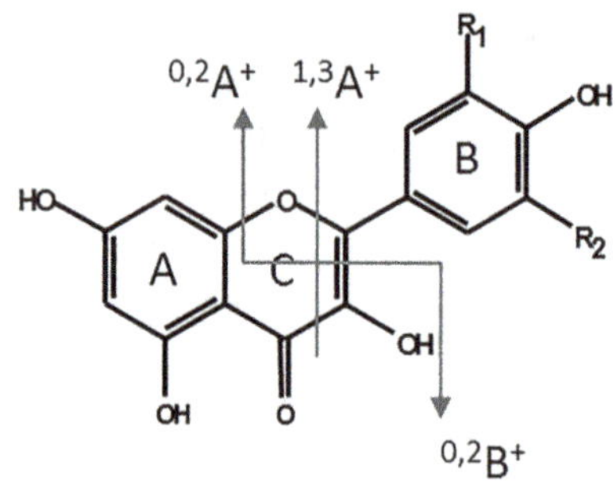

Anthocyanidin	*m/z*	R_1	R_2
Delphinidin	303	OH	OH
Cyanidin	287	OH	H
Petunidin	317	OCH_3	OH
Pelargonidin	271	H	H
Peonidin	301	OCH_3	H
Malvidin	331	OCH_3	OCH_3

Flavonol	*m/z*	R_1	R_2
Myricetin	319	OH	OH
Quercetin	303	OH	H
Laricitrin	333	OCH_3	OH
Kaempferol	287	H	H
Isorhamnetin	317	OCH_3	H
Syringetin	347	OCH_3	OCH_3

Fig. 1 Structures and *m/z* ratios of the main anthocyanidins (left) and flavonol aglycones (right) detected in wines. The main fragment ions resulting from their fragmentation in MS^3 analysis are also indicated in the structures. (Adapted from Ma et al. and Oliveira et al. [9, 10])

different fragments ions can be formed depending on the substitution of B-ring (Figs. 1, 2, and 3). Use the results of the MS^3 analysis to differentiate between the grape native anthocyanins and flavonols whose aglycones show the same *m/z* (*see* **Note 6** and Figs. 1 and 2).

3. To identify the substituent in position 3 in grape native anthocyanins and in flavonols take into account the neutral loss observed from the molecular ion in *full mass spectra* to the aglycone in *MS^2 spectra*: in anthocyanins and flavonols, the loss of 162 amu can be interpreted as the loss of one glucose moiety (or galactose moiety, in the case of flavonols), the loss of 176 (in the case of flavonols), as the loss of a glucuronic acid moiety and the losses of 204, 308 and 324 amu (in anthocyanins) can be interpreted as the losses of one glucose moiety acylated with acetic, *p*-coumaric or caffeic acids, respectively. These acids are linked to the glucose moiety in position 6′, and in the fragmentation conditions usually employed for anthocyanins, the linkage is rarely broken during fragmentation, being the glucose and acid lost together [2–4] (*see* **Note 7**).

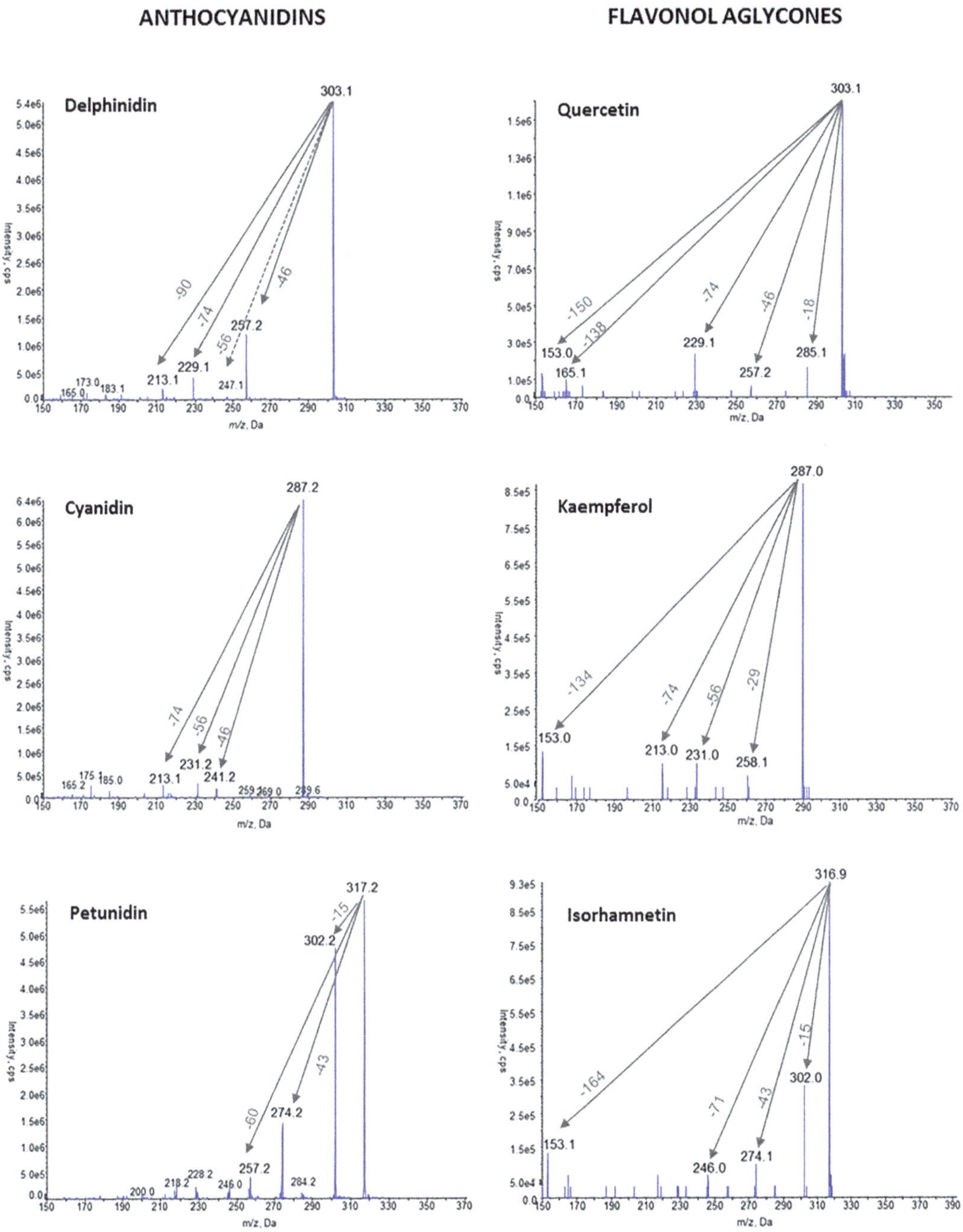

Fig. 2 Comparison of the fragmentation patterns of anthocyanidins (left) and flavonol aglycones (right) sharing the same *m/z* ratios (in positive ion mode: delphinidin and quercetin: 303; cyanidin and kaempferol: 287; petunidin and isorhamnetin: 317). The neutral losses from the aglycone to the different fragment ions are indicated by arrows and numbers

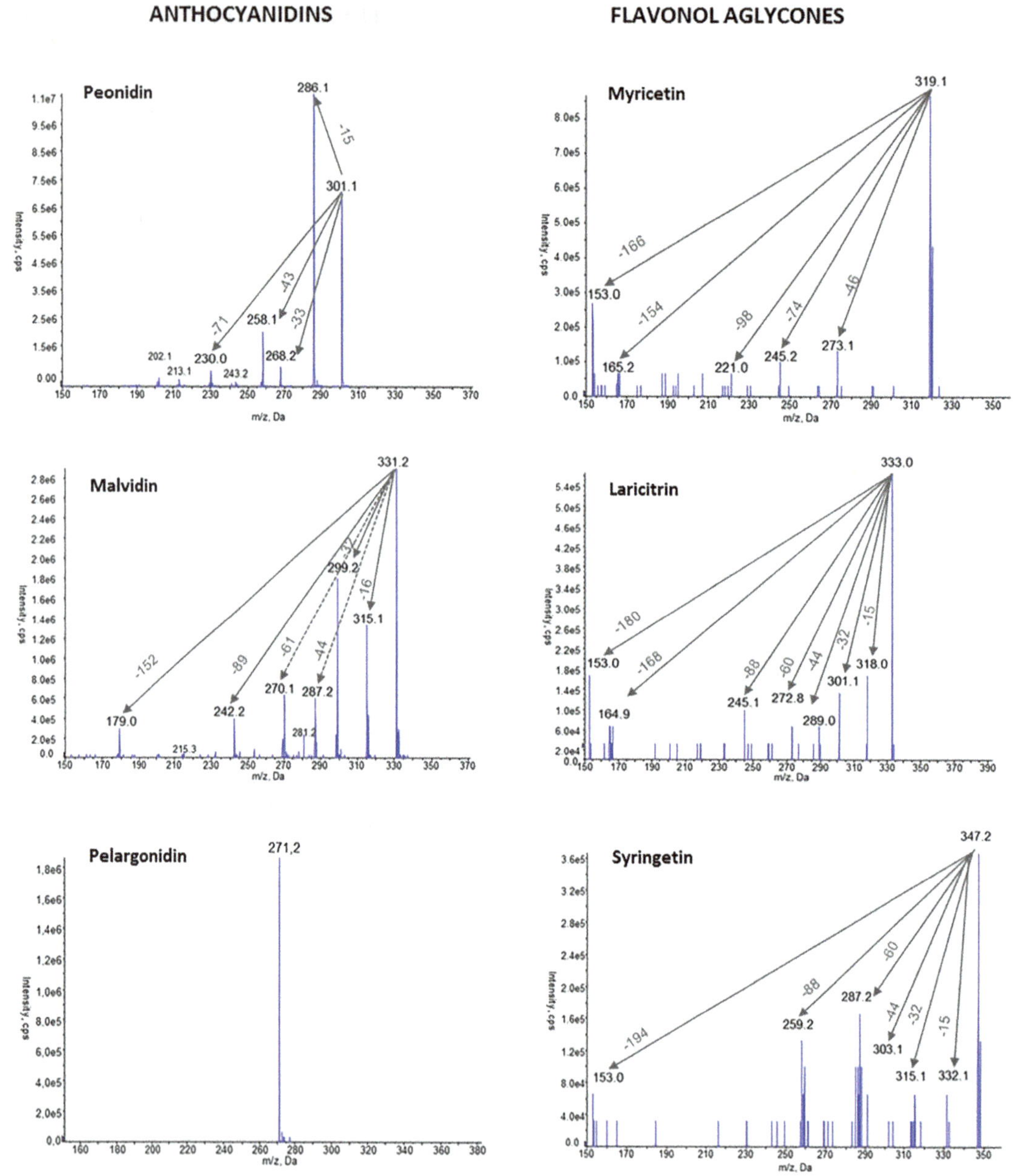

Fig. 3 Fragmentation patterns of the rest of anthocyanidins (left) and flavonol aglycones (right) that do not show the same *m/z*. The neutral losses from the aglycone to the different fragment ions are indicated by arrows and numbers

4. For the identification of the different types of anthocyanin-derived pigments that can be detected in wine samples, take into account these indications:

   • In most cases, as in the case of grape native anthocyanins, the most abundant ion in MS^2 is originated from the loss of the substituent in position 3. Thus, depending on the neutral loss, the anthocyanin-derived compound could be classified as 3-*O*-glucoside, 3-*O*-acetylglucoside, 3-*O*-*p*-coumaroylglucoside or 3-*O*-caffeoylglucoside.

- In the case of *pyranoanthocyanins* [2, 3], the most abundant ion in MS^2 corresponds to the aglycone, and from its *m/z*, the nature of the pyranoanthocyanin can be inferred. The aglycones of A-type vitisins (10-carboxypyranoanthocyanins) and B-type vitisins (formed after cycloaddition of acetaldehyde) show 68 and 24 additional amu in relation to the anthocyanidin from which they derive. In the case of the pyranoanthocyanins derived from hydroxycinnamic acids, the aglycones are 116, 132, or 146 amu greater than the corresponding anthocyanidin depending on the type of hydroxycinnamic acid involved in the synthesis (*p*-coumaric, caffeic, or ferulic acids, respectively) (*see* **Note 8**).
- In *Flavanol-Anthocyanin Direct Condensation Products* (*F-A*$^+$), the *m/z* of the molecular ion is the sum of the molecular mass of the anthocyanin from which it is formed and the molecular mass of the flavanol with 2 amu less (for example, the dimer catechin-malvidin 3-*O*-glucoside shows a molecular ion at *m/z* at $781 = 493 + 290 - 2$). In the MS^2 spectra, the main fragment ion is again the aglycone, and the neutral losses (162, 204, 308, or 324 amu) will be helpful to identify the *substituent in position 3 of the anthocyanidin* (glucose, acetylglucose, *p*-coumaroylglucose, or caffeoylglucose, respectively). Localize in MS^2 and/or MS^3 spectra the "diagnosis" fragment ion for determining the *anthocyanidin* involved in F-A$^+$: it is the one showing the mass of the anthocyanidin and 42 additional amu. In addition, in some cases, a small fragment ion corresponding to anthocyanidin can be observed, but not always (the interflavanic bond, the linkage between position 4 of the flavanol and position 8 of the anthocyanidin and rarely position 6, is stronger than glucosidic bond and is rarely broken). The "diagnosis" fragment ion to determine the identity of the *flavanol* is that resulting from the retro Diels-Alder fission (RDA) in the C-ring of the flavanol (originated from loss of 152 amu when the flavanol involved is (epi)catechin or 168 amu, when it is (epi)gallocatechin) [2–4]. Direct condensation products between flavanols and dimeric anthocyanins (F-A-A$^+$) have been also detected in wines, and mass spectrometry has been a very useful technique to elucidate the identities of the flavanols and anthocyanins involved in these oligomeric structures (*see* **Note 9**).
- In the case of *Flavonol-Anthocyanin Acetaldehyde-mediated Condensation Products* (*F-et-A*$^+$) [2–4], the aglycone is not

the most abundant ion in the MS^2 spectra. Instead, the major product, ion always shows 26 additional amu in relation to the native anthocyanidin and can be considered the "diagnosis" fragment ion to determine the ***anthocyanidin***. In addition, this "diagnosis" fragment ion is further fragmented in MS^3 analysis into fragment ions originated from the same neutral losses as those observed in the fragmentation of the native anthocyanidin (neutral losses shown in Figs. 2 and 3, left). Deduce the *type of flavonol* from the second most abundant ion in MS^2 spectrum since it is formed from a neutral loss of either 290 amu for a (epi)-catechin moiety or 306 amu for a (epi)gallocatechin one. Additionally, a minor fragment ion can appear in the MS^2 spectrum originated after RDA of the flavanol, which can be confirmative of the identity proposed for the flavanol. Infer, as in previous types of pigments, the nature of *substituent in position 3* of the anthocyanidin from the neutral loss between the third most abundant ion in MS^2 spectrum and the molecular ion. This third most abundant ion is the aglycone, which would only contain the flavanol linked to the anthocyanidin through the methyl-methine bridge (*see* **Note 10**).

3.2 Flavan-3-ol Characterization by HPLC-MRM-MS

3.2.1 Sample Preparation

1. Fractionate the wine sample for flavan-3-ols purification by SPE: *(i)* The cartridge must be conditioned with methanol and water (2 mL of each solvent) before using it, *(ii)* load the acidified and diluted sample (2 mL, *see* **Note 11**) onto the cartridge, *(iii)* cleanup the sample with water (4 mL) in order to remove non-phenolic compounds, such as sugars, *(iv)* elute flavan-3-ols (along with flavonols and phenolic acids) with methanol (8 mL), and (v), concentrate this fraction until dryness under vacuum and re-dissolve it using water (0.5 mL) [5].
2. Add to the purified sample, prior to the chromatographic separation, chlorogenic acid as internal standard (final concentration 0.025 mg/mL) to correct the variability in mass spectrometry (*see* **Note 12**).

3.2.2 Flavan-3-ol analysis

1. Perform chromatographic analysis by using C18 columns with superficial porous, core-shell particles and formic acid aqueous solution (0.1% *v/v*) and acetonitrile as solvents [6].
2. Identify flavan-3-ols by means of mass spectrometry (both full mass analysis and collisionally induced fragmentation patterns of $[M+H]^+$ protonated ions) to characterize the degree of polymerization of the flavan-3-ol and the number of (epi)-gallocatechins/(epi)catechins in the flavan-3-ol structure (*see* Table 1 and *see* **Note 13**).

Table 1
MRM transitions that can be used to quantify the most abundant flavan-3-ols in wine

Flavan-3-ol structure	Protonated ion $[M+H]^+$ (*m/z*)	MRM transitions (protonated ion-fragment ion)
(Epi)catechins	291	291-139
Epicatechin gallate	443	291-139
Procyanidin dimers	579	579-289 and 579-291
Procyanidin monogalloylated dimers	731	731-289 and 731-291
Procyanidin digalloylated dimers	883	883-441
Procyanidin trimers	867	867-577 and 867-579
Procyanidin monogalloylated trimers	1019	1019-579, 1019-729 and 1019-731
Procyanidin digalloylated trimers	1171	1171-441
Procyanidin tetramers	1155	1155-577 and 1155-579
Procyanidin monogalloylated tetramers	1307	1307-577 and 1307-579
Procyanidin digalloylated tetramers	1459	1459-729
Procyanidin pentamers	1443	1443-577 and 1443-579
Procyanidin monogalloylated pentamers	1595	1595-729 and 1595-731
(epi)gallocatechins	307	307-139
Epigallocatechin gallate	459	459-139
Prodelphinidin dimers (*pure*)	611	611-305 and 611-307
Prodelphinidin dimers (*mixed*)	595	595-289 and 595-305
Prodelphinidin trimers (*mixed*)	883	883-593, 883-579,and 883-577
Prodelphinidin trimers (*double mixed*)	899	899-609 and 899-593

3. Stablish the MS-MS transitions in positive ion mode for the quantification of the different flavan-3-ol structures, so a multiple reaction monitoring (MRM) of MS-MS transitions method is built on the basis of their protonated ion and the main fragments ions obtained in MS-MS (*see* Table 1) [6].

   3.1 Select the transition "protonated ion *m/z*"-139, based on the retro-Diels Alder reaction (RDA) fission, for the quantification of all monomeric flavan-3-ols (galloylated or not).

3.2 Select the transitions based on the quinone methide (QM) fission for the rest of flavan-3-ols, considering the possibilities of galloylation or B-ring hydroxylation, when necessary. Take into account that it is recommended to use more than a single transition (Table 1) to quantify each type of flavan-3-ol, since this allows increasing the sensitivity and, therefore, decreases the quantification limits. Most of the flavan-3-ol oligomers can be monitored by using two different transitions, excepting procyanidin monogalloylated trimers and prodelphinidin mixed trimers (i.e., prodelphinidins trimers including one (epi)gallocatechin unit and two (epi)catechin units), which need three different MS-MS transition to allow monitoring all the possible isomeric flavan-3-ols showing the same protonated ion (*see* Table 1 and **Note 14**).

4. Determine the signal corresponding to all the transitions set up for each compound and calculate the sum of all of them to quantify each flavan-3-ol (*see* **Note 15**).
5. Correct the variability of the signal by using the signal corresponding to the internal standard employed, whose MS-MS transition must be monitored too. Quantification is performed from the ratio between the flavan-3-ol signal (the sum of all the transitions set up for the corresponding flavan-3-ol) and the internal standard signal (r ratio), since this r ratio is the one that is directly related to flavan-3-ol concentration with the highest precision and accuracy.

3.3 Analysis of Flavonoids by MALDI-TOF

3.3.1 Sample Cleanup (see **Note 16**)

1. Evaporate the wine sample under vacuum and dissolve it in acidic water.
2. The SPE (solid phase extraction) tC18 cartridge is preconditioned (activated with methanol and rinsed with water).
3. Load the acidified wine sample onto the cartridge (it is important to not saturate the cartridge, *see* **Note 17**).
4. Clean the sample with acidified water (acetic acid 1% *v/v*) to remove salts, proteins, and other polar compounds.
5. Elute phenolic compounds with methanol.
6. Concentrate the wine's phenolic extract under vacuum at 32 °C and freeze-dried for further analysis.

3.3.2 MALDI-TOF Analysis

1. Prepare an aqueous solution of 70% (*v/v*) acetone by mixing 700 mL of acetone and 300 mL of water.
2. Prepare a matrix solution: dissolve 100 mg of tIAA (trans-3-indolacrylic acid) matrix in 2 mL of the aqueous solution of 70% (*v/v*) acetone (*see* **Note 18**)

3. Dissolve the wine phenolic extract (obtained after SPE) in an aqueous solution of 70% (*v/v*) acetone in a concentration of 5 mg/mL.
4. Mix 500 μL of the matrix solution with 500 μL of the sample solution, add 10 mg of Dowex 50X8-400 cation exchange resin for deionization.
5. Apply two microliters from each sample (the mixture of the matrix and the resin) onto a stainless-steel target plate and fully air-dried (sample spot).
6. Prepare for each assay three sample spot replicates to analyze them [7].
7. Perform MALDI-TOF analysis. Conditions are: Laser power is adjusted to 55%, the detector gain 4.0× (2410 V), and the mass range between 200 and 3500. Positive ion mode is used for the detection of anthocyanins and negative ion mode is used for flavanols (*see* **Notes 19** and **20**).

4 Notes

1. Usual dilutions range from 1/2 up to 1/5 depending on the color of the wine and still allow the detection of minor compounds. However, as these minor compounds usually co-elute with more abundant compounds, it is difficult to obtain their fragmentation pattern. To overcome this problem, you can program the software of the mass spectrometer to perform a "Dynamic Exclusion" of the major compounds. With this option, the most abundant ions are temporarily excluded from the fragmentation, thus allowing the fragmentation of less abundant ones. Setting the dynamic exclusion of the major compounds after three cycles would be enough to obtain the fragmentation pattern of major and minor compounds co-eluting.
2. Optimize the HPLC conditions to allow a good separation of anthocyanins and anthocyanin-derived compounds and a good separation of flavonols. C-18 reversed-phase columns combined with a gradient of an acidic aqueous eluent and an organic solvent (for instance: TFA 0.1% and acetonitrile, respectively) are adequate for wine analysis
3. Additionally, chromatograms can be recorded at 505 nm to easily localize pyranoanthocyanins by overlapping the chromatogram at 520 nm with that at 505 nm.
4. Calibration curves for anthocyanins and flavonols should be prepared by dissolving the corresponding standard, malvidin 3-*O*-glucoside or quercetin 3-*O*-glucoside, in acidified water or ultrapure water, respectively. From seven to nine different levels

of concentration should be prepared for each compound and analyzed under the same conditions as samples. Standard concentration should range between 0.01 and 100 mg/L in the case of malvidin 3-*O*-glucoside, and between 0.05 and 100 mg/L in the case of quercetin 3-*O*-glucoside. Moreover, to increase precision, the five lowest concentration levels should be used separately to build a calibration curve to quantify low levels of compounds.

5. We have observed that in the mass conditions optimized for anthocyanins, flavonol glycosides are more easily broken than anthocyanins during ionization. Consequently, in order to localize minor flavonols, it is advisable to obtain an extracted ion chromatogram (XIC) at the *m/z* ratios of all of the flavonol aglycones and at the *m/z* ratios of the suspected flavonol glycosides (according to elution order and UV spectrum).

6. The *m/z* ratios of some anthocyanidins and some flavonol aglycones are the same (in positive ion mode: delphinidin and quercetin, *m/z* 303; cyanidin and kaempferol, *m/z* 287; petunidin and isorhamnetin, *m/z* 317). In addition, flavonols can be substituted at position 3 with glucose, just as anthocyanins. Consequently, the *m/z* ratios of the glucosides of both types of compounds will be also identical. If mass analysis is performed coupled to HPLC-DAD, then, the elution order and, above all, the UV-vis spectra will allow the assignment of the identity. If these data are not available, data supplied by mass spectrometer can be helpful for their identification. First, in the conditions usually employed for anthocyanin analysis in wines, flavonol glycosides are more easily fragmented than anthocyanins during ionization, and for this reason, a greater proportion of flavonol aglycone is observed in full mass spectra in relation to anthocyanidins. Second, although the *m/z* rations of the aglycones are identical in some cases, their fragmentations differ (Figs. 1 and 2). The most helpful fragment ion is that at *m/z* 153, present in the fragmentation of all of the flavonol aglycones and not in those of the anthocyanidins (Figs. 2 and 3). This ion (ion $^{1,3}A^{+}$) is formed after the breaking of linkages 1 and 3 in C ring (RDA); Fig. 1), which cannot occur in anthocyanidins in the conditions usually employed due to the existence of conjugated double bonds in these positions [8, 9].

7. Wines made from grapes other than *Vitis vinifera* can show anthocyanidin 3,5-*O*-diglucosides. With the improved sensitivity of mass spectrometry techniques, these compounds have also been detected in small amounts in wines made with *Vitis vinifera* grapes, probably appearing in that species by the use of American rootstocks. Diglucosides of a given anthocyanidin share the *m/z* ratio with the caffeoylmonoglucosides of that anthocyanidin, since the molecular masses of glucose and

caffeic acid are the same. In a combined reversed phase-HPLC-DAD-MS^n analysis, the elution order and different features of their UV-vis spectra would allow their differentiation (for the same anthocyanidin, the diglucoside elutes earlier than the caffeoylglucoside; in the UV spectra of 3-*O*-caffeoylglucosides there is an increased absorption band from 260 to 380 nm, with maxima at 280 and 330 nm and the UV-vis spectra of 3,5-*O*-diglucosides the hump around 440 nm existing in 3-*O*-monoglucosides disappears). In addition, these compounds can be clearly differentiated by their mass fragmentation pattern. In the case of caffeoylglucosides, caffeoyl and glucosyl moieties are lost together (loss of 324 amu), originating a single fragment ion in their MS^2 spectra that corresponds to the anthocyanidin. In contrast, in the MS^2 spectra of 3,5-*O*-diglucosides, an additional fragment ion corresponding to the monoglucoside occurs. This ion can be formed either by the loss of the glucose moiety in 3-*O*-position or by the loss of that located in 5-*O*.

8. The combination of different anthocyanidins and different acids to form hydroxyphenyl- derivatives can originate aglycones with the same *m/z* (for example, the pyranoanthocyanidin derived from peonidin and vinylcatechol and the pyranoanthocyanidin derived from petunidin and vinylphenol show both a signal at *m/z* 595). Similarly, 10-carboxypyranopeonidin 3-*O*-glucoside (A-type vitisin of peonidin) and 10-methylpyranomalvidin 3-*O*-glucoside show the same *m/z* (531), as well as 10-methylpyranopetunidin 3-*O*-glucoside and pyranomalvidin 3-*O*-glucoside (B-type vitisin of malvidin), which share the same *m/z* ratio (517). In these cases, differentiation can be done above all from their UV-vis spectra, but also on the basis of the fragmentation of the aglycone in MS^3 analysis, since the losses observed are the same as those observed in the fragmentation of the native anthocyanidins (Figs. 2, 3 left and 4).

9. Fragmentation of dimeric anthocyanins (A-A$^+$) and direct condensation products between them and flavanols (F-A-A$^+$): In these dimeric and trimeric structures, the fragmentation occurs similarly to that occurring in F-A$^+$ dimers, with fragment ions originated from Retro Diels-Alder fission (RDA)and/or from Heterocyclic Ring Fission (HRF) in the upper units, from the losses of one or two glucose moieties (acylated or not) or from the cleavage of the interflavanic linkage between the flavanol and the upper unit of the dimeric anthocyanin in the case of F-A-A$^+$. A large number of fragment ions can be formed from F-A-A$^+$ since many of these fragmentations can occur simultaneously in different parts of the oligomeric structure. However, as in F-A$^+$ dimers, there are "diagnosis" fragment ions that can

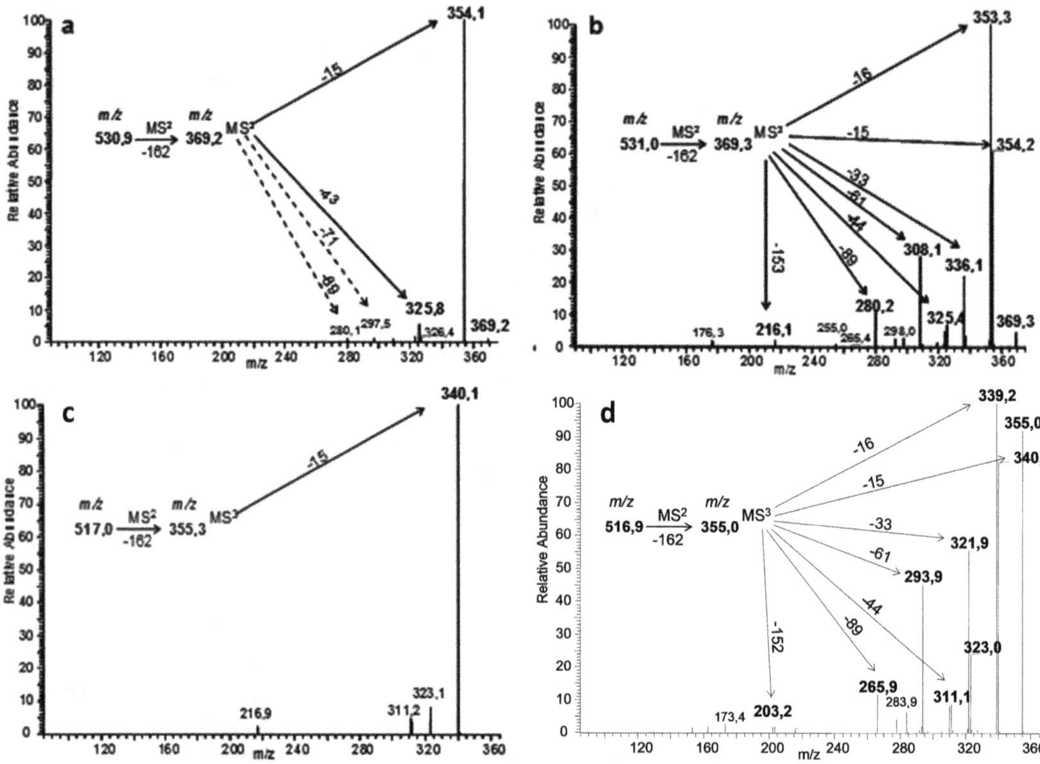

Fig. 4 Examples of fragmentation patterns of pyranoanthocyanins showing the same *m/z*, but deriving from different aglycones and belonging to different pyranoanthocyanin types. a and b: Comparison between 10-carboxypyranopeonidin 3-*O*-glucoside (A-type vitisin of peonidin) and 10-methylpyranomalvidin 3-*O*-glucoside, which both show a molecular ion at *m/z* 531 and an aglycone at *m/z* 369. c and d: Comparison between 10-methylpyranopetunidin 3-*O*-glucoside and pyranomalvidin 3-*O*-glucoside (B-type vitisin of malvidin), which share the same *m/z* ratio (517) and aglycone (*m/z* 355)

inform about the nature of the flavanol, the nature of the anthocyanidin located in the middle position and the nature of the anthocyanidin located in the lower position of the oligomer [10].

10. Differentiation between flavanol *epi*-isomers in F-A$^+$ and F-et-A$^+$ condensation products. The differentiation of *epi*-isomers only from the information supplied by mass spectrometry is difficult given the similarity in the fragmentation of isomers and the low concentration of some of them, which make sometimes difficult the obtaining of MS^3 spectra. Thus, elution order and relative proportions of the isomers are helpful for the assignment of the identity. In the case of F-A$^+$ dimers, those containing catechin or gallocatechin elute earlier than the corresponding compounds containing epicatechin or epigallocatechin. In the case of F-et-A$^+$ dimers, two isomers can be formed for each pair of anthocyanin-flavanol [11, 12]. Consequently, the same *m/z* ratio might correspond to four different

compounds (two *epi*-isomers and two diasteroisomers). The synthesis of these compounds from malvidin 3-*O*-glucoside, acetaldehyde and catechin or epicatechin in studies carried out in our laboratory [2, 3, 11, 12] has allowed the determination of the proportions between diasteroisomers (the second one is formed in a greater proportion) and the elution order of all the four compounds (epicatechin-first diasteroisomer, catechin-first diasteroisomer, catechin-second diasteroisomer, and epicatechin-second diasteroisomer).

11. Sample is diluted with 0.1 M HCl (1:1) in order to put all pigments in their cationic form so they can be removed during SPE.
12. 10 μL of a solution of chlorogenic acid of 1 mg/mL are mixed to 390 μL of the re-dissolved fraction to reach a final concentration of chlorogenic acid of 0.025 mg/mL without an important dilution of the sample.
13. The collisionally induced fragmentation patterns of $[M+H]^+$ protonated ions of monomeric flavan-3-ols show fragment ions derived from four main losses [13]. First, the loss of a molecule of water (18 amu) gives rise to a fragment at *m/z* 273 and 289, for (E)C and (E)GC, respectively. The main fragmentation occurs through the retro-Diels-Alder (RDA) fission of C ring, which implies the loss of a neutral fragment of 152 amu for (E)C or 168 for (E)GC leading, in both cases, to a fragment at *m/z* 139. Also, the heterocyclic ring fission (HRF) of C ring implies the loss of a neutral fragment of 126 amu, giving rise to fragments at *m/z* 165 or 181 that can lose a molecule of water and give fragments at *m/z* 147 or 165 for (E)C and (E)GC, respectively. Besides, benzofuran-forming (BFF) fission of the initial molecule can give rise to two ionized fragments, one at *m/z* 123 or 139 (for (E)C or (E)GC, respectively) and one at *m/z* 169 for both types of monomeric flavan-3-ols. In addition to this fragmentation pattern, the oligomeric and polymeric flavan-3-ols undergo the quinone methide (QM) fission implying the rupture of the interflavanic bond. Thus, for instance, the main fragmentation in the case of B-type procyanidin dimers ($[M+H]^+$ at *m/z* 579) implies the QM fission giving rise to two ionized fragments at *m/z* 289 and *m/z* 291, which will appear along with all the fragments corresponding to the fragmentation pattern described above. Also, the chromatographic behavior, that is, the relative retention time, must be considered to differentiate between flavan-3-ols of the same molecular weight and, even like this, most of the isomers cannot be unequivocally identified [14].
14. The conditions for multiple reaction monitoring (MRM) analysis have to be optimized by using the different flavan-3-ol

types, regarding their degree of polymerization, galloylation, and hydroxylation of B-ring. The main parameter that should be adapted to each structure is the collision energy used to fragment the protonated ion, giving rise to the selected fragment ion. In the case of (epi)catechins and procyanidins, the collision energies use to be lower for non-galloylated monomeric, dimeric, and trimeric flavan-3-ols and higher for more polymerized structures. However, when galloylated, the trimeric flavan-3-ols need similar collision energy than more polymerized structures and higher than the dimeric and monomeric ones. On the contrary, in the case of (epi)gallocatechins and prodelphinidins, the highest collision energies are used for the monomeric structures (both galloylated or not) [6].

15. For a given protonated ion, not all the fragments selected in the MS-MS transitions can be detected, since it depends on the position of the sub-units to which the galloyl group is linked to, when the flavan-3-ol is a galloylated derivative and on the position of the (epi)gallocatechin units in the mixed and double mixed prodelphinidins. This way, in that case, it is possible that one compound shows signal just at a determined MS-MS transition, which will be used for quantification. However, in most cases, flavan-3-ols will show signal at all transitions set up, although with a different intensity depending on their structure.
16. It is important to do a sample cleanup, in order to eliminate other families of compounds that could cause interference with the mass signal, increasing background noise and diminishing the quality of the *m/z* signal.
17. To avoid the saturation of the solid phase, the wine sample should be in a small volume and high concentration and the color should not diffuse over one-third of the cartridge when depositing the sample.
18. For polyphenol analysis, the most commonly used matrices are tIAA (trans-3-indolacrylic acid) but DHB (2,5-dihydroxybenzoic acid) can be also employed.
19. Proanthocyanidins (or tannins) can be detected in positive ion mode or negative ion mode (*see* Fig. 5 as an example). When negative ion mode is used, a simpler spectrum is obtained, since the phenolic compounds are usually detected in positive ion mode as sodium and/or potassium adducts. For instance, in Fig. 5 there is only one mass signal per compound in a series of galloylated and non-galloylated dimers and trimers, the most intense *m/z* corresponds to the galloylated trimer (*m/z* 1017).

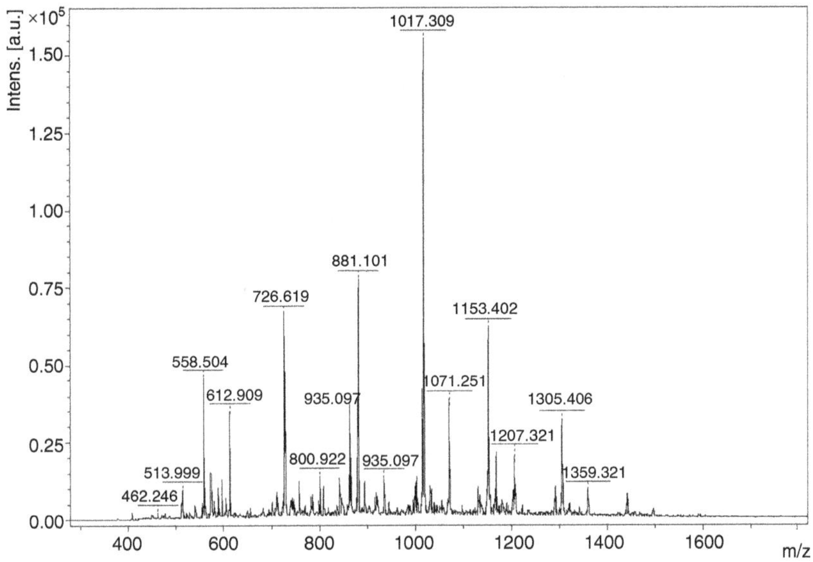

Fig. 5 MALDI-TOF analysis in negative ion mode of a wine tannin fraction

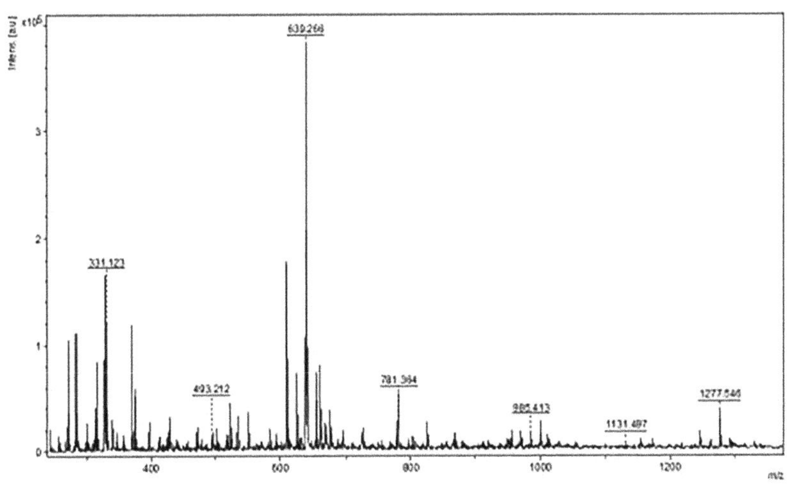

Fig. 6 MALDI-TOF analysis in positive ion mode of a wine pomace extract

20. Anthocyanins are best detected in the positive ion mode because they are naturally occurring cations. See Fig. 6 as an example of a MALDI-TOF analysis of anthocyanins, in which wine pomace fraction is analyzed, allowing the determination of malvidin 3-p-coumaroylglucoside (*m/z* 639) and other native (anthocyanin dimers) and derived anthocyanins (*m/z* 609 and *m/z* 781).

Acknowledgments

This research was financially supported by Grant PID2021-127126OB-C21 funded by MCIN/AEI/10.13039/501100011033 and by "ERDF A way of making Europe." Thanks are also due to Junta de Castilla y León-FEDER Programme (Strategic Research Programs for Units of Excellence Project Ref. CLU-2O18-04). E. Salas thanks Spanish MIU for the María Zambrano contract.

References

1. Alcalde-Eon C, Pérez-Mestre C, Ferreras-Charro R, Rivero FJ, Heredia FJ, Escribano-Bailón MT (2019) Addition of mannoproteins and/or sedes during winemaking and their effects on pigment composition and color stability. J Agric Food Chem 67:4031–4042
2. Alcalde-Eon C, Escribano-Bailón MT, Santos-Buelga C, Rivas-Gonzalo JC (2004) Separation of pyranoanthocyanins from red wine by column chromatography. Anal Chim Acta 513: 305–318
3. Alcalde-Eon C, Escribano-Bailón MT, Santos-Buelga C, Rivas-Gonzalo JC (2006) Changes in the detailed pigment composition of red wine during maturity and ageing. A comprehensive study. Anal Chim Acta 563:238–254
4. Alcalde-Eon C, Boido E, Carrau F, Dellacassa E, Rivas-Gonzalo JC (2006) Pigment profiles in monovarietal wines produced in Uruguay. Am J Enol Vitic 57(4):449–459
5. González-Manzano S, Santos-Buelga C, Pérez-Alonso JJ, Rivas-Gonzalo JC, Escribano-Bailón MT (2006) Characterization of the mean degree of polymerization of proanthocyanidins in red wines using liquid chromatography-mass spectrometry (LC-MS). J Agric Food Chem 54:4326–4332
6. García-Estévez I, Alcalde-Eon C, Escribano-Bailón MT (2017) Flavanol quantification of grapes via multiple reaction monitoring mass spectrometry. Application to differentiation among clones of *Vitis vinifera* L. cv. Rufete grapes. J Agric Food Chem 65:6359–6368
7. Reed JD, Krueger CG, Vestling MM (2005) MALDI-TOF mass spectrometry of oligomeric food polyphenols. Phytochem 66:2248–2263
8. Ma YL, Li QM, Van den Heuvel H, Claeys M (1997) Characterization of flavone and flavonol aglycones by collision-induced dissociation tandem mass spectrometry. Rapid Commun Mass Spectrom 11:1357–1364
9. Oliveira MC, Esperança P, Almoster Ferreira MA (2001) Characterisation of anthocyanidins by electrospray ionisation and collision-induced dissociation tandem mass spectrometry. Rapid Commun Mass Spectrom 15:1525–1532
10. Alcalde-Eon C, Escribano-Bailón MT, Santos-Buelga C, Rivas-Gonzalo JC (2007) Identification of dimeric anthocyanins and new oligomeric pigments in red wine by means of HPLC-DAD-ESI/MS^n. J Mass Spectrom 42: 735–748
11. Rivas-Gonzalo JC, Bravo-Haro S, Santos-Buelga C (1995) Detection of compounds formed through the reaction of malvidin 3-monoglucoside and catechin in the presence of acetaldehyde. J Agric Food Chem 43:1444–1449
12. Francia-Aricha EM, Guerra MT, Rivas-Gonzalo JC, Santos-Buelga C (1997) New anthocyanin pigments formed after condensation with flavanols. J Agric Food Chem 45: 9349–9352
13. Li HJ, Deinzer ML (2007) Tandem mass spectrometry for sequencing proanthocyanidins. Anal Chem 79:1739–1748
14. Teixeira N, Azevedo J, Mateus N, de Freitas V (2016) Proanthocyanidin screening by LC–ESI-MS of Portuguese red wines made with teinturier grapes. Food Chem 190:300–307

Chapter 5

Quantification of Proteins in White and Rosé Wines

Richard Marchal

Abstract

The Bradford method is used worldwide in biochemistry to quantify proteins. Proteins might react with a dye, the Coomassie Brilliant Blue (CBB), to form blue complexes measured at 595 nM. The Bradford method is the most commonly used method in oenology for must and wine protein quantification because of its rapidity, reproducibility, accuracy, and ease of implementation. Nevertheless, different interferences may falsify the quantification of protein because of interactions between nonprotein compounds and the dye reagent. Thus, ultrafiltration with 3–10 kDa molecular weight cut-off (MWCO) membrane showed that nonprotein compounds largely react with the Coomassie Brilliant Blue (CBB) and can produce an $A_{595\ nM}$ equal to 30–90% of the value estimated by direct quantification. Ethanol and phenolic compounds are the main responsible for these interferences. Moreover, both bentonite and vegetable charcoal treatments partially eliminate proteins and phenolics. In this chapter a modified Bradford method involving the measurement of $A_{595\ nM}$ before and after ultrafiltration (3–10 kDa MWCO) and taken in consideration the global interference with the CBB, will be presented, allowing us for a more accurate measure of the wine protein content.

Key words Protein estimation, Bradford assay, White wine, Interferences, Oenological treatments

1 Introduction

Precipitation of wine proteins results in the formation of hazes and deposits during wine aging or storing under unsuitable conditions [1, 2]. It has also been shown that proteins actively participate in the formation and the stabilization of the foam in sparkling wines [3]. On the other hand, it has been observed that yeast mannoproteins play an important part in tartaric precipitation and sparkling wine foam [4, 5]. The importance of protein properties in wine behavior is the reason why these macromolecules have been the focus of many studies, with a particular interest in their quantification.

The need to use a reliable method seems evident after having read the literature results. In fact, reported protein contents are extremely variable, depending on the grape type, the vintage, and

María Ángeles Pozo-Bayón and Carolina Muñoz González (eds.), *Wine Analysis and Testing Techniques*,
Methods and Protocols in Food Science, https://doi.org/10.1007/978-1-0716-3650-3_5,

the vinification method. This variability in protein content is also due to sample preparation and to the meaning that is given to the term "protein": nitrogenous fraction excluded by gel-permeation chromatography, precipitable (by ethanol, trichloroacetic acid, or ammonium sulfate), isolated by SDS-PAGE after CBB staining, or soluble in the wine. The employed methods to quantify wine-soluble proteins are also variable: Biuret, Lowry, A_{280}, (direct) Bradford, Smith, Kjeldahl, SDS-PAGE, or by a proteomic approach. Since Hsu and Heatherbell experiments [6], the Bradford method [7] is probably the most widely used in oenology for white and rosé wine protein quantification [8–11], even if the potassium dodecyl sulfate/bicinchoninic acid (KDS/BCA) is sometimes preferred in the case of haze risk studies [12]. The several advantages of this method—rapidity, reproducibility, accuracy, sensitivity, low cost, and above all usable in direct dosage—have led to general use according to many authors [9–11].

With this Bradford method, white and rosé wine protein concentrations vary from 1 mg/L [8] to 220 mg/L [11]. This incredible variability in concentration is explained by whether or not interferences are considered.

In fact, interferences are inherent to all biochemical protein estimation methods. Thus, for example, phenolic compounds, oligopeptides (MW $<$ 10 kDa), and ethanol strongly react with CBB [8].

In this chapter, we present a corrected Bradford method [8] for a much better estimation of soluble proteins in musts and white/rosé wines, even if it is known that there is no perfect method. For this, we discuss the advantages and the limits of the modified Bradford method used in oenology.

The principle on which this Bradford corrected method is based is a global estimation of the interference ($A_{595\ nM}$ of the wine ultrafiltrate) which will be subtracted from the value obtained with the direct dosage ($A_{595\ nM}$ read after mixing the wine and the CBB dye reagent).

The problem of interference of non-protein compounds with CBB can be illustrated with two examples of oenological treatments: vegetable carbon and bentonite. The values in red (Fig. 1) indicate the "apparent" protein content without correction for global interference. The values in blue indicate the protein levels actually present in the wine. This graph shows the errors made with the direct Bradford assay, especially when trying to understand the impact of bentonite fining or carbon treatment on the total soluble protein content of a wine.

Figure 1 is given as an example of a white wine made from Pinot noir grapes (however, the example would be quite similar with Chardonnay). The values in red represent the apparent protein content, in mg/L equivalent BSA, following the quantification of proteins by using the Bradford method in a direct assay. These

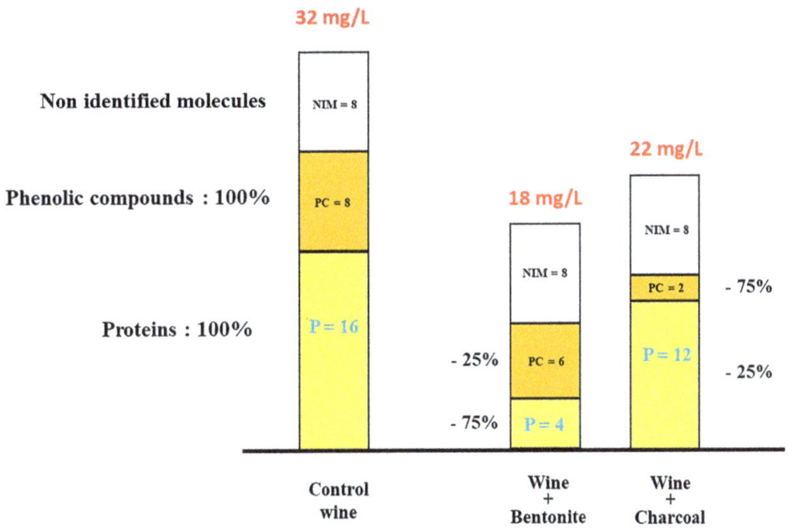

Fig. 1 Wine proteins quantification with the direct Bradford method: illustration of the variation in overall interference as a function of oenological treatments

values (in red) consider interferences as this is a direct assay with the wine in its entirety, without subtracting anything. For this control wine, the A_{595} corresponds to 32 mg/L eq BSA. But, really, this wine contains only 16 mg/L of protein. Phenolic compounds react with CBB to the extent of 25% of the total A_{595} or 8 mg/L eq BSA. In addition, many other unidentified compounds (among which ethanol plays a major role) also react at 25% of the total A_{595}, that is, also 8 mg/L eq BSA, even though these compounds are not proteins.

When treated with bentonite, the wine protein content will fall by 75% in this example, to just 4 mg/L. As bentonite does not specifically eliminate proteins, phenolic compounds are also adsorbed and their concentration will fall by 25%. Interference from phenolic compounds will therefore fall from 8 to 6 mg/L eq BSA. In the end, this wine after bentonite fining will react with the Bradford reagent to give an A_{595} equal to 18 mg/L eq BSA. If the control wine is treated with vegetable charcoal, the protein content will fall from 16 to 12 mg/L, a reduction of 25%. On the other hand, the phenolic compound content will fall by 75%, as vegetable charcoal is very refined for these molecules. Interference from phenolic compounds, using the direct Bradford assay, will fall from 8 to 2 mg/L eq BSA. In the end, using the direct Bradford assay this wine will react with the CBB reagent to give an A_{595} equal to 22 mg/L eq BSA. Using the direct assay, the apparent difference between the bentonite-treated wine (18 mg/L eq BSA) and the charcoal-treated wine (22 mg/L eq BSA) is 4 mg/L; the charcoal-treated wine appears to contain 22% ($22/18 = 1.22$) more protein than the bentonite-treated wine. But really, if we look at the real protein content (in blue), we observe that the charcoal-treated wine (12 mg/L actual protein) contains 3 times more protein than the bentonite-treated wine (12 mg/L actual protein) (*see* **Notel**).

This example shows the extent to which the Bradford method gives an erroneous result when used in direct dosage. It is therefore essential to correct for overall interference in order to be able to interpret the impact of an oenological treatment or cultivation practice.

The Bradford method proposed in this chapter has been adopted by the OIV as the reference method for the quantification of proteins in white wines (OIV-OENO RESOLUTION 625-2021).

This modified Bradford method is probably the most accurate for quantifying proteins in wines. However, it does have its limitations, as do all methods of quantifying proteins when they are glycosylated (*see* **Note 2**).

2 Materials

- 1 mL plastic micro-cuvettes. Disposable or in glass. Light path: 1 cm.
- Bradford reagent (Bio-Rad—it is very easy to use and the shelf life is over a year) (*see* **Note 3**).
- Distilled water.
- Ultrafiltration units (to receive 1–2 mL of sample) with 3, 5, 8, or 10 kDa-MWCO (*see* **Note 4**).
- Spectrophotometer.
- Bovine Serum Albumin (BSA).
- Filtered sample of must or wine (0.45 or 0.22 μm). Many kinds of filters can be used. Nevertheless, it is preferable to use filters with a low affinity for proteins, such as nitrocellulose filters.

3 Methods

3.1 Standard Curve

Protein content should be calculated with regard to bovine serum albumin (BSA Fraction purified, without lipids) standard curve (0–20 mg/L).

To prepare the standard curve, Mix: 400 μL water +400 μL BSA solution (5–20 mg/L) + 200 μL Bradford reagent. The BSA must be prepared in a buffer (20–50 mM, with tartaric, phosphate, or citric acid, for example, and adjust the pH to a value close to that of a wine—generally between 3.1 and 3.7, according to the region and the style of the wine) to facilitate its solubilization. For the protein-free control sample, mix 400 μL water, 400 μL buffer, and

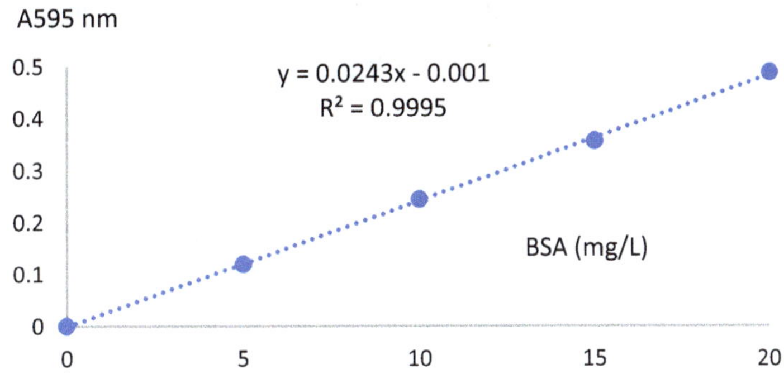

Fig. 2 Example of BSA calibration curve

200 μL of Bradford reagent. The absorbance A_{595} of the protein-free control (the blank without BSA) should be close to 0.480–0.520. This value (average of 3 measurements) must be subtracted from all measurements. For example, if the control has an $A_{595} = 0.500$ and the wine has an $A_{595} = 0.825$, the wine protein absorbance A_{595} is 0.825–0.500 = 0.325.

For the calibration curve (Fig. 2), $A_{595\ nM}$ should be measured after 4–8 min of incubation ($A_{595\ nM}$ max) at room temperature (*see* **Note 5**). Each value corresponds to the average of three measurements. All standard deviations are less than 2%. The protein content of the must or wine will be expressed in mg/L equivalent BSA (mg/L eq. BSA).

3.2 Determination of Wine/must Soluble Protein

The Bradford method is employed as follows: 200 μL Bradford reagent (Coomassie Blue Brilliant, CBB), 400 μL distilled water, and 400 μL sample. As musts and wines are already buffered media, there is no need to use a buffer; distilled water is perfectly suitable.

The blue coloration should be measured at 595 nM using a spectrophotometer after 45–60 min of contact at room temperature when the color is stable. The must and the wines can be diluted if necessary so that the $A_{595\ nM}$ readings remain below 1 (200 μL Bradford reagent +600 μL mL distilled water +200 μL mL sample). This is always the case for wine with a high protein content, such as Muscat (Moscatel) or Gewurztraminer.

3.3 Calculation to Correct Interferences

1. Estimation of wine (or must) reactivity with the CBB dye reagent by the direct Bradford method = blue coloration of the wine measured at 595 nM,

   Wine Direct Estimation: W_{DE}

2. Ultrafiltration of the wine (or must) with a 3 or 10 kDa MWCO membrane and estimation of the ultrafiltrate reactivity with the dye reagent = blue coloration of the ultrafiltrate (UF) measured at 595 nM,

   Ultrafiltrate Reactivity Estimation: UF_{RE}

3. Determination of the (corrected) Wine Protein Content W_{PC}, W_{PC} (mg/L eq. BSA) = W_{DE} − UF_{RE}/slope of the BSA curve.

   For example: $W_{DE} = 0.874$ at 595 nM.

   $UF_{RE} = 0.526$ at 595 nM.

   BSA calibration curve: $y = 0.0243x - 0.001$.

   Slope ($\Delta y/\Delta x$) = 0.0243 (A_{595} for a solution at 1 mg/L).

   $W_{PC} = (0.874–0.526)/0.0243 = 14.3$ mg/L eq. BSA.

4 Notes

1. When we follow the decrease of proteins in a wine following a bentonite treatment at different doses, we can clearly see the effect of the doses applied: the protein content always decreases as the dose of bentonite increases. In this case, the decrease in protein content is estimated quite correctly with the Bradford method if one considers that most of the proteins are adsorbed by the clay particles in the same proportions. However, if proteins such as MPs are less adsorbed (because of their high glycosylation level), then the calculation of the decrease in total protein has a margin of error.
2. CBB only reacts with the polyaminoacid part of proteins. However, in must and wine, there are many glycoproteins, both of plant (such as arabinogalactan proteins or grape berry invertase) and yeast origin (essentially Mannoproteins, noted MPs). As a result, the protein content estimated with the corrected Bradford method does not consider the complete molecular structures. For yeast MPs, with an average level of glycosylation around 95%, the polyaminoacid part represents only 5% of the weight of the molecule. For 100 mg of MPs, the Bradford method will give a value of only 5 mg. The limitations of this method can be seen here and illustrated with the following example. When aging on yeast lees, the wine MPs content is increased, without changing the plant protein content. With the Bradford method (even corrected), there is a very slight increase in protein content (only a few mg), while the released MPs content can reach 50–100 mg/L, depending on the aging on Lees method.
3. Bio-Rad Bradford reagent is used in many publications, but it is not the only brand. It is also possible to prepare the CBB reagent in the laboratory, but the product is not stable and the repeatability is so as good as with the Bio-Rad reagent.

4. By pure convention in biochemistry, proteins are compounds with a molar mass greater than 10 kDa. However, the compounds involved in foam stabilization or in the wine haze do not fit within these arbitrary limits, and one may wish to estimate fractions of different molecular weights: above 3, 5, 8, 10, 20, 30, 50, and 100 kDa for example. So, if a 3, 5, or 8 kDa MWCO membrane is used for wine ultrafiltration, it is preferable to speak of high molecular weight nitrogen (HMWN) compounds (W_{HMWN}).
5. Do not wait for any longer, as intense blue particles will appear due to the precipitation of BSA in an acidic medium.

References

1. Pasquier G, Feilhes C, Dufourcq T, Geffroy O (2021) Potential contribution of climate change to the protein haze of white wines from the French Southwest Region. Foods 10:1355. https://doi.org/10.3390/foods10061355
2. Cosme F, Fernandes C, Ribeiro T, Filipe-Ribeiro L, Nunes FM (2020) White wine protein instability: mechanism, quality control and technological alternatives for wine stabilisation – an overview. Beverages 6:19. https://doi.org/10.3390/beverages6010019
3. Kemp B, Condé B, Jégou S, Howell K, Vasserot Y, Marchal R (2019) Chemical compounds and mechanisms involved in the formation and stabilization of foam in sparkling wines. Crit Rev Food Sci Nutr 59(13): 2072–2094. https://doi.org/10.1080/10408398.2018.1437535
4. Moine-Ledoux V, Perrin A, Paladin I, Dubourdieu D (1997) First result of tartaric stabilization by adding mannoproteins (Mannostab™). J Int Sci Vigne Vin 31:23–31
5. Marchal R, Salmon T, Gonzalez R, Kemp B, Vrigneau C, Williams P, Doco T (2020) Impact of Botrytis cinerea contamination on the characteristics and foamability of yeast macromolecules released during the alcoholic fermentation of a model grape juice. Molecules 25(3):472. https://doi.org/10.3390/molecules25030472
6. Hsu JC, Heatherbell DA, Flores JH, Watson BT (1987) Heat-unstable proteins in grape juice and wine. II. Characterization and removal by ultrafiltration. Am J Enol Vitic 38: 17–22
7. Bradford MM (1976) A rapid and sensitive method for the quantitation of microgram quantities of protein utilizing the principle of protein-dye binding. Anal Biochem 72:248–254
8. Marchal R, Seguin V, Maujean A (1997) Quantification of interferences in the direct measurement of proteins in wines from the Champagne Region using the bradford method. Am J Enol Vitic 48:303–309
9. De Bruijn J, Loyola C, Arumí JL, Martínez J (2014) Effect of non-protein factors on heat stability of Chilean Sauvignon Blanc wines. Chil J Agric Res 74(4):490–496
10. Hung WF, Harrison R, Morton JD, Trought MCT, Frost A (2014) Protein concentration and bentonite requirement in Marlborough Sauvignon Blanc wines. Aust J Grape Wine Res 20:41–50
11. Saracino F, Brinco J, Gago D, Gomes da Silva M, Boavida Ferreira R, Ricardo da Silva J, Chagas R, Ferreira LM (2021) DCMC as a promising alternative to bentonite in white wine stabilization. impact on protein stability and wine aromatic fraction. Molecules 26: 6188. https://doi.org/10.3390/molecules26206188
12. Gazzola D, Vincenzi S, Pasini G, Lomolino G, Curioni A (2015) Advantages of the KDS/BCA assay over the bradford assay for protein quantification in white wine and grape juice. Am J Enol Vitic 66:227–233. https://doi.org/10.5344/ajev.2014.14076

Chapter 6

White and Rosé Wine Haze Risk (WHR) Estimation

Richard Marchal, Thomas Salmon, Jacques Emmanuel Barbier, and Bertrand Robillard

Abstract

The appearance of cloudiness in a white or rosé wine is a phenomenon that occurs in most countries around the world. To avoid this, winemakers treat their wines, in most cases with bentonite, and very often with excessive doses. But how are these inappropriate treatments chosen to obtain wine colloidal stability? They are based on a heat test in which the wine is heated to 80 °C for 30 min to 2 h, generally. At this temperature of 80 °C, many of the thermostable wine proteins that remain soluble during a heatwave (when a bottle of wine can rise to 35–42 °C) are insolubilized. It is therefore necessary to change the heat test. A study conducted in different wine-producing regions, with many grape varieties, was carried out over 4 years. The results showed that the most appropriate test was to heat the wine to 50 °C for 90 min. The cloudiness induced is very closely correlated with the cloudiness observed during a heat wave while maintaining a safety margin for winemakers. With this laboratory test, in which the wine is heated for 90 min at 50 °C, the doses of bentonite are considerably reduced compared with the doses deemed necessary on the basis of a test at 80 °C. For wines that present a low risk according to the 80 °C test, the 50 °C test shows that treatment with bentonite can even be abandoned. The use of this 50 °C test therefore radically changes the approach to protein stabilization of white and rosé wines.

Key words Haze risk estimation, Turbidity, White wine, Rosé wine, Heat test, Oenological treatments

1 Introduction

Cloudiness in a bottle of white or rosé wine is unquestionably perceived as a defect for the consumers. It is therefore essential to ensure the stability of the wine clarity after bottling, whatever the conditions encountered during transport and throughout the storage of the wine, including during heatwaves. To decide which oenological treatment to use, the Wine Haze Risk (WHR) is generally (although not exclusively) estimated by a heat test in which the wine is heated at 80 °C for 30 min to 6 h (for review see reference [1]) (*see* **Note 1**). Then, the cloudiness (turbidity)

María Ángeles Pozo-Bayón and Carolina Muñoz González (eds.), *Wine Analysis and Testing Techniques*, Methods and Protocols in Food Science, https://doi.org/10.1007/978-1-0716-3650-3_6,

generated by this heat treatment is measured after cooling the sample with a turbidimeter.

These conditions are very different considering the temperatures to which a wine may be subjected, even in heatwave conditions. This significant difference in temperature between the heat tests carried out in the laboratory and the temperature that a white/rosé wine is really exposed to has led us to reconsider the relevance or accuracy of this 80 °C test. This approach is also in line with the numerous observations that wine producers and oenology laboratories have been reporting for years that can be summarized as follows: the 80 °C test overestimates the WHR estimation [1–3]. It's also clear that summers will get hotter and hotter, and that global warming is a reality. According to a recent study, the WHR is higher in years with hot weather [4]. So which test should we use?

1.1 What Is Turbidity and What Is a Turbidimeter?

In enology (and for liquids in general), turbidity is a parameter defining the degree to which wine loses its transparency because of suspended solids particles. A turbidimeter is a portable or in situ instrument to measure suspended particles present in the wine (or the grape juice) through the light scattering. It measures the suspended particles with a light beam (source) and a light detector receiving the light at 90° from the original beam. Thus, the light reflected by the particles suspended is a function of the particle density. In addition, the light reflected by the particles depends on their shape, color, and reflectivity. For this reason, two wines with the same turbidity can have different particle compositions, in number and size. A turbidimeter is calibrated with a known particulate material (formazine generally).

1.2 A Realistic WHR Test

To answer this question "which test should we use?," which is complicated due to the high variability in colloidal composition of white and rosé wines, different heat tests were compared. They mimicked heat wave temperatures (35, 38, 42, and 46 °C), the duration of which can be counted in days, with quicker laboratory heat tests at 50 °C and 80 °C which take place over a few tens of minutes [5–9]. These studies allowed us the possibility: (1) to determine the overestimation of the WHR with laboratory tests and (2) to select the test more correlated with reality.

These studies were carried out over 4 years, with white/rosé wines from different French and Spanish regions (from cool to hot climates), and produced with different grape varieties (Sauvignon, Moscatel/Muscat, Gewurztraminer, Pinot noir, Riesling, Chardonnay, Mourvèdre/Tempranillo).

Twenty-one tests were carried out with a large number of wines. An example of a white Sauvignon wine is given in Fig. 1 [5].

The above-mentioned study and results shown in Fig. 1, drive three main conclusions:

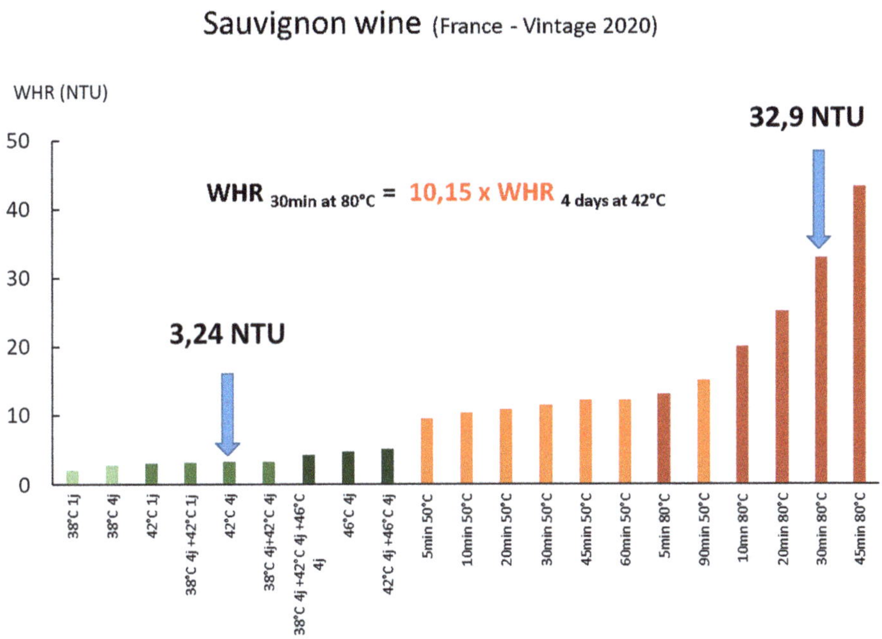

Fig. 1 Measurement of the cloudiness (NTU) that appears in a Sauvignon white wine subjected to 21 different heat tests. Green, the heat test temperatures correspond to possible heat waves (reality). Orange, turbidity after heating at 50 °C (realistic lab test). Red, turbidity after heating to 80 °C (excessive lab test). After heating for 30 min at 80 °C, turbidity was 10.15 times higher than the turbidity observed after a 4-day heatwave at 42 °C (*see* **Note 2**)

(i) The 80 °C test overestimates the WHR by a factor of 10–13 if we take as reference a 4-day heat test at 42 °C (which is already a temperature rarely reached by a bottled wine), and even more if we take as reference a 4-day heat test at 38 °C.

(ii) The test at 50 °C, with heating maintained for 60–90 min, gives results much closer to the heat wave realities to which a bottle of white wine may be subjected than the 80 °C test.

(iii) As observed in Table 1 hereafter, the Pearson correlation coefficients (PCC) between the heatwave tests and the 50 °C tests are significantly higher than the PCC reflecting the relationships between the heatwave tests and the 80 °C tests [7].

For these three main reasons, the 50 °C heat test is much more suitable for estimating the WHR of wine than the 80 °C test.

A realistic estimation of the WHR with a 50 °C heat test allows for the implementation of an appropriate and realistic oenological treatment. For many wines, this clearly means a reduction (or even elimination) of the amount of bentonite used for colloidal stabilization. It is also possible to reconsider or develop technical alternatives [5, 10–18] that are considered insufficient if we refer to the 80 °C heat test, but quite efficient if we refer to the 50 °C test (Fig. 2).

Table 1
Relationship between laboratory heat tests and heat wave reality. Pearson correlation coefficients for all of the tests compare 2 by 2

	5 min 80 °C	30 min 80 °C	60 min 80 °C	30 min 50 °C	1 h 50 °C	2 h 50 °C	1 day 35 °C	4 days 35 °C	1 day 42 °C	4 days 42 °C	1 day 46 °C	4 days 46 °C
5 min 80°C	1											
30 min 80 °C	0.993	1										
60 min 80 °C	0.958	0.979	1									
30 min 50 °C	0.980	0.985	0.942	1								
1 h 50 °C	0.979	0.986	0.947	0.999	1							
2 h 50 °C	0.951	0.968	0.931	0.992	0.993	1						
1d 35 °C	0.930	0.950	0.911	0.980	0.980	0.993	1					
4 days 35 °C	0.949	0.963	0.924	0.989	0.989	0.994	0.996	1				
1 day 42 °C	0.935	0.955	0.916	0.984	0.985	0.997	0.997	0.993	1			
4 days 42 °C	0.889	0.917	0.878	0.951	0.951	0.975	0.980	0.962	0.987	1		
1 day 46 °C	0.834	0.866	0.827	0.901	0.902	0.932	0.936	0.906	0.949	0.986	1	
4 days 46 °C	0.812	0.848	0.814	0.883	0.885	0.920	0.923	0.892	0.937	0.979	0.998	1

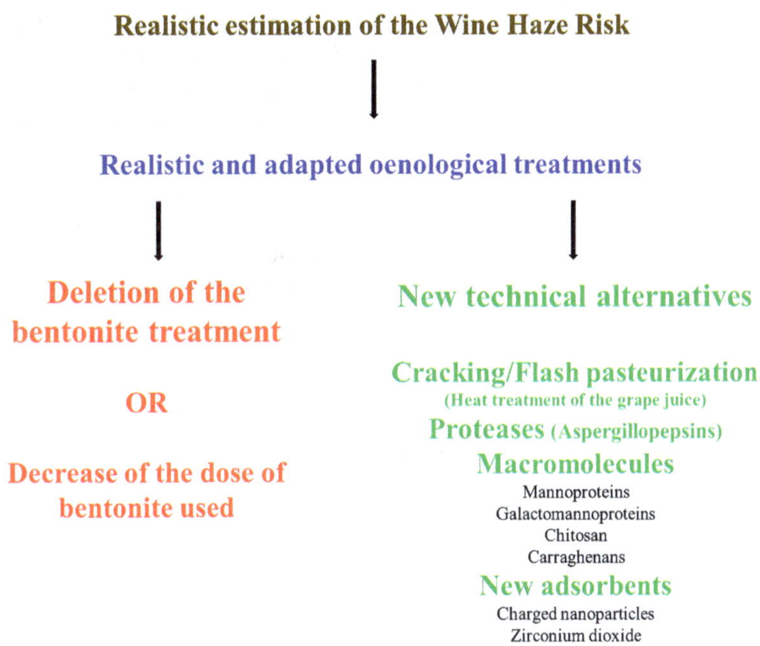

Fig. 2 Possible oenological practices to eliminate the risk of cloudiness appearing in the bottle. These possibilities depend on the test used to assess the WHR

The following sections explain the methodology for a more reliable WHR estimation.

2 Materials

- Laboratory centrifuge (to reduce the sample turbidity and to avoid a quick clogging of the filter).
- Plastic syringe.
- Filters with 0.45 μm cutting membranes (0.22 and 0.65 μm are also useable).
- Thermostatic bath.
- Turbidimeter with calibration kit.

3 Methods

In principle, determining the WHR is technically very simple: you take a wine with no particles visible to a person (naked eye—visual observation). To be rigorous, the wine must be filtered through a calibrated filter (centrifugation is too random and does not give accurate results). This clear wine is then heated (respecting a time/temperature pairing). After cooling, the cloudiness in the wine is measured using laboratory equipment (turbidimeter), which once again is more accurate than the human eye. The appearance of cloudiness after a heat test means that the wine may be cloudy following a heatwave, and that oenological treatment is necessary before bottling. The main steps of the test are explained as follows.

3.1 Turbidimeter Calibration

The turbidity calibration standards provided by the manufacturer are simple solutions for calibrating and validating turbidimeters that have a 0–1000 NTU range (portable) or a 0–4000 NTU range (in-situ equipment). Turbidity calibration standards are generally prepared from NIST traceable primary standard reference material and come supplied with a certificate of analysis.

3.2 Determination of the WHR

- Filter the white and rosé wines using 0.45 μm cutting membranes before the heat tests (0.22 and 0.65 μm filters can be also used).
- Place the wines in glass tubes (as used for example in microbiology) equipped with a screw cap and a seal to prevent any loss of liquid during the heating (8–20 mL of wine/tube, according to the turbidimeter used). The turbidity measured after the heat test does not depend on the volume of wine. It is recommended to do the heat test in triplicate for each wine, even if the results are highly repeatable.
- For the heat wave conditions (choose a precise temperature, between 35 and 42 °C; the temperature is chosen according to producers' requirements), the wines are placed in a culture chamber or a thermostatic bath.
- For the 50 °C heat test, the tubes should be immersed in the thermostatic bath to allow a faster temperature rise due to the much shorter heating time.
- At the end of the heat treatment, the wines should be left at laboratory temperature for at least 2 h or until the following day. Results will be approximately the same or equal.
- Measure the turbidity of the wines with a portable or in situ turbidimeter after calibration (*see* **Note 3**). This turbidity corresponds to the WHR.

4 Notes

1. As proteins become insoluble when the temperature of the wine is increased, it makes sense to use a heat test to assess the risk of cloudiness in the bottle. For this reason, we do not recommend tests based on acidification of the wine or the addition of tannins (these two types of tests give aberrant results).

2. The haze observed by the 50 °C heat test still shows an overestimation of the WHR, which can be considered as a safety margin for wines subject to extreme conditions.
3. The turbidity measured after the heat test should not be corrected with the turbidity measured before the heat test. Many publications show that the turbidity read after heating/cooling is subtracted from the initial turbidity. For example, if a wine has a turbidity of 2.4 NTU before heating and 4.8 after heating, many authors consider the value of 4.8 – 2.4 = 2.4 NTU. This value corresponds to the turbidity generated by heating. However, what matters is what consumers see, and their eye does not subtract. That means they see a turbidity of 4.8 NTU. Therefore, this value matters. For this reason, the preparation (laboratory filtration) of the wines for the heat test should correspond as closely as possible to the filtration that the wine will undergo before bottling.

References

1. Cosme F, Fernandes C, Ribeiro T, Filipe-Ribeiro L, Nunes FM (2020) White wine protein instability: mechanism, quality control and technological alternatives for wine stabilisation – an overview. Beverages 6:19. https://doi.org/10.3390/beverages6010019
2. Meistermann E. Stabilisation protéique des vins blancs et rosés (2010). www.infowine.com. Revue internet de viticulture et œnologie, N°10/2
3. Vernhet A, Meistermann E, Cottereau P, Charrier F, Chemardin P, Poncet-Legrand C (2020) Wine thermosensitive proteins adsorb first and better on bentonite during fining: practical implications and proposition of alternative heat tests. J Agric Food Chem 68(47): 13450–13458
4. Pasquier G, Feilhes C, Dufourcq T, Geffroy O (2021) Potential contribution of climate change to the protein haze of white wines from the French Southwest Region. Foods 10:1355. https://doi.org/10.3390/foods10061355
5. Marchal R, Salmon T, Pudelko M, Robillard B (2022) Casse protéique des vins blancs: evaluation du risque et gestion du traitement à la bentonite. Revue des Œnologues 182:51–55
6. Marchal R. Revisión de la metodología para la determinación de la estabilidad proteica en vinos blancos. Congreso Internacional ACE de la Enología y el Cava 2020, 23 de octubre de 2020. Video of the conference: https://youtu.be/Y6e4tAdnipw. Portail de la revue Acenología: http://www.acenologia.com/index.htm
7. Marchal R, Lecomte M, Salmon T, Robillard B Which heat test really represents the haze risk of a white sauvignon wine from the French Loire Valley? Macrowine 2021, June 21–23, Verona, Italy
8. Marchal R, Salmon T, Pudelko M, Robillard B. Quiebra proteica en los vinos blancos: evaluación del riesgo y gestión del tratamiento con bentonita. Acenología, on line the 23rd of february 2022. https://www.acenologia.com/quiebra_proteica_en_vinos_blancos_evaluacion_riesgo_y_gestion_tratamiento_bentonita/
9. Marchal R., Salmon T., Lecomte M., Robillard B. Which heat test can realistically estimate white wine haze risk? Wine Active Compounds (WAC) conference 2022, June 29–July 1, Dijon, France
10. Saracino F, Brinco J, Gago D, Gomes da Silva M, Boavida Ferreira R, Ricardo-da-Silva J, Chagas R, and Ferreira L. M. DCMC as a promising alternative to bentonite in white wine stabilization. Impact on protein stability and wine aromatic fraction. Molecules 2021, 26, 6188. https://doi.org/10.3390/molecules26206188
11. Salazar FN, Achaerandio I, Labbé MA, Güell C, López F (2006) Comparative study of protein stabilisation in white wine using zirconia and bentonite: physiochemical and wine sensory analysis. J Agric Food Chem 54:9955–9958
12. Marangon M, Lucchetta M, Waters EJ (2011) Protein stabilisation of white wines using

zirconium dioxide enclosed in a metallic cage. Aust J Grape Wine Res 17:28–35

13. Mercurio M, Mercurio V, Gennaro B, Gennaro M, Grifra C, Langella A, Morra V (2010) Natural zeolites and white wines from Campania region (Southern Italy): a new contribution for solving somo oenological problems. Period Miner 79:95–112
14. Mierczynska-Vasilev A, Wahono SK, Smith P-A, Bindon K, Vasilev K (2019) Using zeolites to protein stabilize white wines. ACS Sustain Chem Eng 7:12240–12247
15. Vincenzi S, Mosconi S, Zoccatelli G, Pellegrina CD, Veneri G, Chignola R, Peruffo A, Curioni A, Rizzi C (2005) Development of a new procedure for protein recovery and quantification in wine. Am J Enol Vitic 56:182–187
16. Colangelo D, Torchio F, De Faveri DM, Milena L (2018) The use of chitosan as alternative to bentonite for wine fining: effects on heat-stability, proteins, organic acids, colour, and volatile compounds in an aromatic white wine. Food Chem 264:301–309
17. Ratnayake S, Stockdale V, Grafton S, Munro P, Robinson AL, Pearson W, McRae JM, Bacic A (2019) Carrageenans as heat stabilisers of white wine. Aust J Grape Wine Res 25:439–450
18. Gonzalez-Ramos D, Cebollero E, Gonzalez R (2008) A recombinant Saccharomyces cerevisiae strain overproducing mannoproteins stabilizes wine against proteins haze. Appl Environ Microb 77:5533–5540

Chapter 7

Determination of Biogenic Amines in Wines

Juan José Rodríguez-Bencomo

Abstract

The analysis of biogenic amines (histamine, putrescine, cadaverine, tyramine, and others) in wine is usually carried out by high-performance liquid chromatography (HPLC) with a derivatization previous step to improve the chromatographic separation and the detectability. Here is described a method based on the derivatization with diethyl ethoxymethylenemalonate (DEEMM), and analysis by HPLC and UV spectrophotometric detection.

Key words Wine, Biogenic amines analysis, Histamine, Putrescine, Cadaverine, Derivatization, Diethyl ethoxymethylenemalonate (DEEMM), High-performance liquid chromatography (HPLC)

1 Introduction

The presence of biogenic amines in fermented food products is usual due to their origin being related to the fermentation process itself and to the enzymatic decarboxylation of amino acids by the microorganisms [1, 2]. In the case of wine, the main source of these amines is the malolactic fermentation that is mainly carried out during red winemaking. The major biogenic amines in wines are histamine, putrescine, tyramine, and cadaverine (see in Fig. 1 the chemical structures of these compounds), and their levels depend on several factors, such as the microorganism strain, must/wine composition and the fermentation conditions [3, 4]. Although nowadays there is not a specific regulation about the levels of biogenic amines in wines, the OIV (International Organization of Vine and Wine) in the "OIV Code of good vitivinicultural practices" has recommended minimizing the presence of biogenic amines in vine-based products [5, 6]. This recommendation is based on the potentially unhealthy effects when biogenic amines are ingested with food products. In the case of histamine, these negative effects can be enhanced by the presence of alcohol, acetaldehyde, and other biogenic amines, due to the enzymatic inhibition

María Ángeles Pozo-Bayón and Carolina Muñoz González (eds.), *Wine Analysis and Testing Techniques*, Methods and Protocols in Food Science, https://doi.org/10.1007/978-1-0716-3650-3_7,

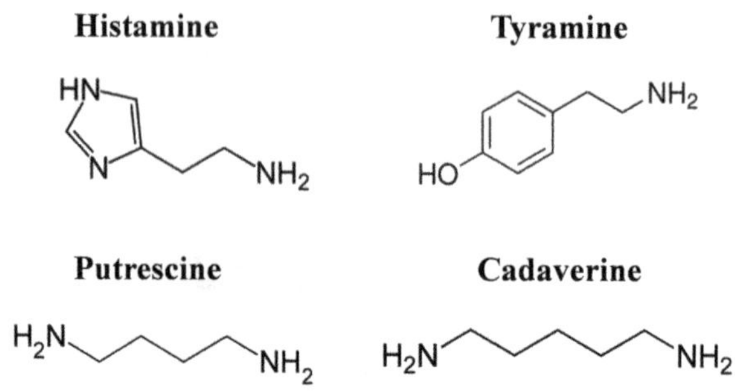

Fig. 1 Chemical structures of the main biogenic amines of wine

of its normal metabolism in humans [2]. Therefore, for monitoring the levels of the biogenic amines in wines will be necessary for accurate and enough sensitive analytical methods according to the usual levels of biogenic amines in wines.

The analysis of biogenic amines in wine samples can be carried out with different separative analytical techniques, such as gas chromatography (GC), high-performance liquid chromatography (HPLC), or capillary electrophoresis (CE). In addition, electrochemical or optical sensors for some specific amines have been developed [7]. Among these options, HPLC is the most usual technique, and, in general, a derivatization step of the amines will be required to improve the chromatographic separation and the detectability. Several derivatization reagents have been used such as dansyl chloride, o-phthalaldehyde (OPA), benzoyl chloride, 6-aminoquinolyl-Nhydroxysuccinimidyl carbamate, 4-chloro-3,5-dinitrobenzotrifluoride, and diethyl ethoxymethylenemalonate (DEEMM) [8]. Among them, DEEMM derivatization has been previously proposed by Alaiz et al. [9] for amino acids derivatization and, more recently, by Gómez-Alonso et al. [10] for the derivatization of amino acids and biogenic amines in wines and beer samples. DEEMM allows a simple derivatization protocol, and the analysis of the amine-DEEMM derivatives can be carried out with common detectors in HPLC (by UV spectrophotometry at $\lambda = 280$ nm).

The OIV in its compendium of methods (edition 2022) [11] includes two methods for biogenic amine analysis in wines: derivatization with DEEMM coupled to HPLC and UV spectrophotometric detection [12], and derivatization with OPA coupled to HPLC and fluorometric detection [13]. This chapter focuses on the derivatization method with DEEMM and HPLC analysis based on the methods of Gómez-Alonso et al., OIV [10, 12], and other published adaptations [14, 15].

2 Materials

2.1 Reagents, Standards, and Solutions

Solutions must be prepared using ultrapure water. All reagents must be of analytical grade or HLPC grade (*see* **Note 1**).

1. Biogenic amines standards (analytes): Histamine [10, 12, 15], methylamine [12], ethylamine [12], tyramine [10, 12, 15], putrescine [10, 12, 15], cadaverine [10, 12, 15], phenylethylamine [10, 12, 15], isoamylamine [10, 12, 15], spermidine [10, 15] tryptamine [10] and spermine [15] (in brackets are indicated the references in which the compound is analyzed).
2. Biogenic amines standard solutions: weight the standard reagents by using an analytical balance. Prepare stock solutions at 1000 mg/L of each amine separately in acid solution (HCl 0.1 M). Store at 4 °C.
3. Prepare a stock amine solution (from the previous separate solutions) in HCl 0.1 M. For example, 100 mg/L of each analyte. The concentrations of this solution must be prepared considering the maximum concentration level of calibration. Store at 4 °C.
4. Calibration solutions: Prepare a synthetic wine solution (12% ethanol, tartaric acid 3.5 g/L, pH = 3.5, adjusted with NaOH 1 M). Prepare the biogenic amines calibration solutions at different levels (usually up to 20 mg/L) in volumetric flasks of 5 mL, by addition of different volumes of the stock amine solution and the synthetic wine. Store at 4 °C.
5. Internal standard (IS): Weigh 2,4,6-Trimethylphenethylamine by using an analytical balance. Prepare a solution of IS at 2000 mg/L in HCl 0.1 M. Store at 4 °C (*see* **Note 2**).
6. Borate Buffer 1 M, pH = 9.0: Weight boric acid and dissolve it in distilled water, adjust the pH with a NaOH (a concentrated solution, for example, 4 M), and finally with NaOH 1 M. If the stored solution presents borate crystals, heat slightly and stir until complete solution. Store at room temperature.
7. Derivatization reagent: diethyl ethoxymethylenemalonate (DEEMM) is used directly (without dilution).

2.2 Analytical Instrument and Chromatographic Column Type

HPLC is equipped with a gradient pump and a UV spectrophotometric detector (or a photodiode array detector). Analysis with mass spectrometer detectors can be also viable.

The chromatographic analysis is carried out in reverse phase mode by using a C18 chromatographic column (suggested characteristics: 250 mm × 4.6 mm and 5 μm of particle size).

2.3 Mobile Phases for HPLC Analysis (See Note 3)

1. Phase A: 2.5 mM acetate buffer (pH = 5.8). To avoid microorganism proliferation, add 0.02% sodium azide or 0.1% Tetrahydrofuran.
2. Phase B: Acetonitrile/Methanol (80/20) (v/v).
3. Filter mobile phases before use through 0.45 μm filter.

3 Methods

3.1 DEEMM Derivatization

1. The derivatization process is carried out in screw cap glass tubes of 5 or 10 mL.
2. Add to each tube: 1.75 mL of borate buffer, 1 mL of wine sample (or synthetic sample of calibrations), 0.75 mL of methanol, 40 μL of internal standard solution, and 30 μL of DEEMM [10, 12]. Shake the mixture.
3. Derivatization reaction is carried out in an ultrasonic bath for 30 min. The derivatization reaction and the reaction product (amine-DEEMM derivative) are presented in Fig. 2 (*see* **Note 4**).
4. Keep tubes at 70 °C in a stove for 2 h to degrade the excess of DEEMM. Wait until room temperature.
5. Filter the sample (0.45 μm filter) and store it in a chromatography vial. DEEMM derivatives are stable at least for a week according to the studies of Gómez Alonso et al. [10].

3.2 HPLC Analysis

1. Injection volume (suggested): 50 μL.
2. UV spectrophotometry detector (or a photodiode array detector): measure of absorbance at $\lambda = 280$ nm.
3. Identification of analytes and IS in the chromatogram by retention time: according to the chromatographic column characteristics and the mobile phases chosen, the chromatographic separation must be optimized by application of an adequate gradient program (*see* **Note 5**). Therefore, the retention times of analytes and internal standard must be determined by injection of the standards separately.

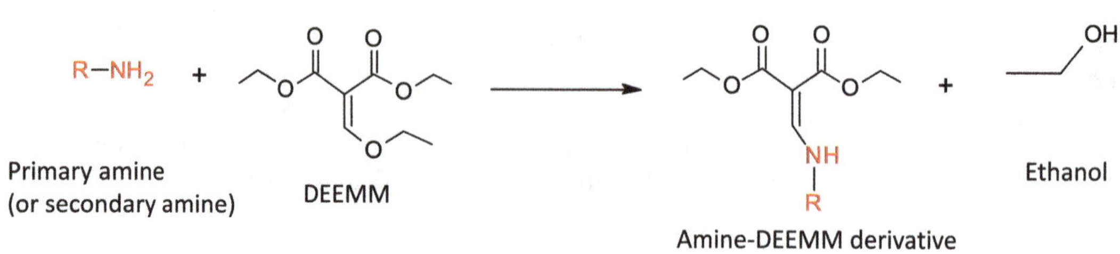

Fig. 2 Derivatization reaction of amines with DEEMM

4. As an example: for the mobile phases proposed in the OIV method [12] (phase A: 2.5 mM sodium acetate; phase B: Acetonitrile/Methanol (80/20), flow rate of 0.9 mL/min and a C18 column (250 mm × 4.6 mm and 5 um of particle size; column temperature at 15 °C), the elution gradient used can be: 0 min 90% A; $t = 5$ min 90% A; $t = 10$ min 83% A; $t = 35$ min 60% A; $t = 43$ min 28% A; $t = 48$ min 18% A; $t = 52$ min 0% A; $t = 57$ min 0% A. In these conditions, the retention time of the last eluted compound (2,4,6-Trimethylphenethylamine (IS)) is around 55 min, being the retention times of the analytes: histamine (≈25 min), methylamine (≈33 min), ethylamine (≈39 min), tyramine (≈41 min), putrescine (≈46 min), cadaverine (≈48 min), phenylethylamine (≈49 min), and isoamylamine (≈50 min) (*see* **Note 6**).
5. Inject samples and calibration curves and introduce a quality control sample (such as a reference sample or a spiked sample with the analytes) in each analysis series.
6. See **Note 7** for a basic protocol for method validation.

3.3 Data Treatment and Results

1. Integrate the chromatograms to obtain the peak area for each compound. Transform the peak area to the relative peak area ($A_{relative}$) by dividing by the internal standard area ($A_{relative}= =Area_{analyte}/Area_{IS}$) for each chromatogram. Working with relative peak area allows correct possible bias due to derivatization or injection.
2. Built the calibration curves ($A_{relative}$ vs. Concentration (mg/L)) and obtain the linear regressions for each analyte. Obtain the limit of detection and quantification of the method by using blank samples and/or calibration parameters.
3. Interpolate the $A_{relative}$ of samples into the linear regressions to obtain the concentration of each analyte.
4. Express the results in mg/L.

4 Notes

1. For the preparation of standards solution of biogenic amines and internal standards work with a mask and/or in a fume hood is suggested due to the toxicity and strong odor of the pure compounds.
2. Other possible internal standards: L-2-aminoadipic acid [10] and n-heptylamine [15].
3. From the results obtained by Rodríguez Bencomo et al. [15], in the case of multiple amino groups in the biogenic amine molecule, multiple reactions could be produced. Thus, histamine, isoamylamine tyramine, and β-phenylethyamine

derivatizates are formed by the reaction with only one DEEMM molecule, however, cadaverine and putrescine react with two DEEMM molecules, and spermine and spermidine with three DEEMM molecules. According to that, reaction of primary and secondary group amine occurs, but steric hindrance could limit the reaction of specific amine groups such as in the case of spermine.

4. These mobile phases are proposed by Gomez Alonso et al. [10] and OIV-OENO 457/2014 [12]. Other authors have used different mobile phases for the chromatographic separation of amines-DEEMM derivatives using mass spectrometer detectors [15, 16], for instance: phase A: 1 mM. ammonium acetate +0.1% formic acid (adjusted to pH = 3.2); and phase B: Acetonitrile.
5. Derivatization with DEEMM also implies the formation of the derivative products of amino acids of the sample, so, in the case of analysis by UV spectrophotometry detection ($\lambda = 280$ nm), the correct separation of the analytes avoiding coelutions must be verified.
6. If a mass spectrometer is used as detector, the elution gradient can be adapted for a shorter analysis time. By using a triple quadrupole detector in ESI + mode, sodium adducts of the molecular ions are observed [15]. Multiple Reaction Monitoring (MRM) transitions for quantification are the following: histamine (304.2 → 258.3), putrescine (451.2 → 405.2), cadaverine (465.2 → 419.1), spermine (735.3 → 416.1), spermidine (678.3 → 348.3), tyramine (330.0 → 284.1), 2-phenylethylamine (314.1 → 268.2), and isoamylamine (280.1 → 162.3).
7. Basic protocol for method validation:
   (a) Evaluate the quality parameters of the complete method: Repeatability and reproducibility, detection and quantification limits, dynamic linear ranges, and recoveries. For quality parameters of the method, see references [10, 12].
   (b) Evaluate the accuracy of results by regular analysis of reference samples or by using the addition standard method.

References

1. Sahu L, Panda SK, Paramithiotis S, Zdolec N, Ray RC (2015) Biogenic amines in fermented foods: overview. In: Montet D, Ray RC (eds) Fermented foods—part I: biochemistry and biotechnology. CRC-Press, Boca Raton, pp 318–332
2. Smit AY, du Toit WJ, du Toit MS (2008) Biogenic amines in wine: understanding the headache. South African J Enol Vitic 29:109–127
3. Moreno-Arribas MV, Polo MC (2009) Amino acids and biogenic amines. In: Moreno-Arribas

MV, Polo MC (eds) Wine chemistry and biochemistry. Springer, New York, pp 163–189
4. Smit I, Pfliehinger M, Binner A, Grossmann M, Horst WJ, Lohnertz O (2014) Nitrogen fertilisation increases biogenic amines and amino acid concentrations in Vitis vinifera var. Riesling musts and wines. J Sci Food Agric 94:2064–2072
5. OIV (2011) OIV code of good vitivinicultural practices in order to minimise the presence of biogenic amines in vine-based products. OIV-CST 369-2011. https://www.oiv.int/public/medias/2766/oiv-cst-369-2011-en.pdf
6. OIV (2020) OIV guide to identify hazards, critical control points and their management in the wine industry. Resolution OIV-OENO 630-2020. https://www.oiv.int/public/medias/7592/oiv-oeno-630-2020-en.pdf
7. Ordoñez JL, Troncoso AM, García-Parrilla MC, Callejón RM (2016) Recent trends in the determination of biogenic amines in fermented beverages. A review. Anal Chim Acta 939:10–25
8. Hernández-Cassou S, Saurina J (2011) Derivatization strategies for the determination of biogenic amines in wines by chromatographic and electrophoretic techniques. J Chromatogr B 879:1270–1281
9. Alaiz M, Navarro JL, Girón J, Vioque E (1992) Amino acid analysis by HPLC after derivatization with diethyl ethoxymethylenemalonate. J Chromatogr 591:181–186
10. Gómez-Alonso S, Hermosín-Gutiérrez I, García-Romero E (2007) Simultaneous HPLC analysis of biogenic amines, amino acids, and ammonium ion as aminoenone derivatives in wine and beer samples. J Agric Food Chem 55:608–613
11. OIV (2022) Compendium of international methods of wine and must analysis. Ed 2022, 1. https://www.oiv.int/public/medias/8613/2022-eng-compendiumvol1.pdf
12. OIV (2014) A method of determination of biogenic amines in wine by high-performance liquid chromatography with photodiode array detection. Resolution OIV-OENO 457/2014
13. OIV (2009) Analysis of biogenic amines in musts and wines HPLC. Resolution OIV-OENO 346/2009
14. Bach B, Le Quere S, Vuchot P, Grinbaum M, Barnavon L (2012) Validation of a method for the analysis of biogenic amines: histamine instability during wine sample storage. Anal Chim Acta 732:114–119
15. Rodríguez-Bencomo JJ, Rigou P, Mattivi F, López F, Mehdi A (2020) Removal of biogenic amines from wines by chemisorption on functionalized silica and effects on other wine components. Sci Reports 10:17279
16. Rebane R, Oldekop ML, Herodes K (2012) Comparison of amino acid derivatization reagents for LC–ESI-MS analysis. Introducing a novel phosphazene-based derivatization reagent. J Chromatogr B 904:99–106

Chapter 8

Determination of Free and Glycosidically Bound Fractions Responsible of Grape Musts Aroma by Solid-Phase Extraction (SPE) and Gas Chromatography-Mass Spectrometry (GC-MS)

Mar Vilanova and José M. Oliveira

Abstract

The analysis of volatile composition in grape musts is an important tool to characterize grape cultivars and to determine the influence of biotic and abiotic stresses in this composition. One of the most common methods used for sample preparation to identify and quantify the volatile compounds is solid-phase extraction (SPE).

The purpose of this work is to describe SPE methodology applied to grape must to identify and quantify a high number of grape aroma compounds in the two fractions, volatile and glycosidically bound, with a reduced quantity of sample and solvents. This methodology was applied by the authors to characterize the major grape cultivars (*Vitis vinifera*) from the North of Iberian Peninsula (Galicia and north Potugal).

Key words GC-MS, Grape must, SPE, Volatile compounds

1 Introduction

Volatile composition of grape must determine the organoleptic characteristics of wines being an informative tool for characterization of grape cultivars. The aroma profile of grape cultivars is attributed to a variety of compounds that can be separated into various classes based on their chemical structure. Moreover, many of these compounds in grapes exist as glycosidically bound precursors [1–3], which are released by acidic and enzymatic hydrolysis during crushing, fermentation, and storage [4]. Therefore, the complete analysis of grape aroma is complex, expensive and it consumes a lot of time.

One of the most common methods used for sample preparation/cleaning up and concentrating the volatile compounds is solid-phase extraction (SPE), which can handle a wide range of

María Ángeles Pozo-Bayón and Carolina Muñoz González (eds.), *Wine Analysis and Testing Techniques*, Methods and Protocols in Food Science, https://doi.org/10.1007/978-1-0716-3650-3_8,

chemical classes and concentrations. This technique has evolved along the years [3, 5–12]. SPE is still the more suitable method for grape aroma compounds analysis because it allows concentration of the analytes down to the microliter scale after starting from a considerable volume of sample [13]. SPE remains the best sample preparation method for higher boiling volatiles, such as monoterpenols, and allows extraction and concentration of large sample volumes with high recovery of the analytes [13]. Moreover, the advantage over other methods is that this method not only permits the extraction of free fraction of aroma but also the glycosidically bound fraction. Glycosidically bound volatile compounds are typically analyzed following isolation of the glycosides on columns, including solid-phase extraction (SPE), filled with a C18-bonded reversed-phase stationary phase [3, 5, 9, 14–16].

In the last years, there has been interest in developing a rapid and direct approaches for analysis of glycosidically bound compounds to quantify the aroma potential of the grape. In this sense, Dungey et al. [17] using high performance liquid chromatography with tandem mass spectrometry measured directly guaiacol glycoconjugates. Automated approaches for SPE isolation have also been proposed [18]. Schneider et al. [19] described a rapid analysis of grape aroma glycoconjugates using Fourier transform infrared spectrometry without the need of any hydrolysis; however, in this case, glycoside purification by SPE was necessary. No rapid methods have been developed, so far, for measuring concentration of individual glycosidically bound aroma compounds without the need for time-consuming sample purification before analysis [20].

Over the past 20 years, headspace solid phase microextraction (HS-SPME) has emerged as a suitable tool for rapid analysis of volatile compounds, but this technique only is useful to extract the free fraction [20].

In this paper, we describe solid-phase extraction (SPE) methodology applied to grape must to extract a high number of grape aroma compounds in the two fractions, volatile and glycosidically bound, with a reduced quantity of sample and solvents. Several additional rapid sample preparation steps are also described. This methodology was applied by the authors to characterize the major grape cultivars (*Vitis vinifera*) from the North of Iberian Peninsula (Galicia and north Potugal) [21, 22].

2 Materials

2.1 Chemicals and Reagents

- Chromatographic grade reagents: ethanol (99.8%), dichloromethane (99.8%), methanol (99.9%), pentane (99.9%), ethyl acetate (99%), and sodium sulfate (99%).
- Pure standards to identify volatile compounds (Table 1).

Table 1
Standards used to identify volatile compounds

Chemical group	Compound	CAS No.	Formula	Molecular weight M/(g/mol)	Purity (> %)
TERPENOLS	Linalool	78-70-6	C10H18O	154.25	97
	α-Terpineol	10482-56-1	C10H18O	154.25	96
	Citronellol	106-22-9	C10H20O	156.27	95
	Nerol	106-25-2	C10H18O	154.25	97
	Geraniol	106-24-1	C10H18O	154.25	98
	Terpinen-4-ol	562-74-3	C10H18O	154.25	92
C_{13}-NORISOPRENOIDS	β-Damascenone	23696-85-7	C13H18O	190.28	nd
	α-Ionone	127-41-3	C13H20O	192.3	90
	β-Ionone	79-77-6	C13H20O	192.3	96
ALCOHOLS	1-buthanol	71-36-3	C4H10O	74.12	99
	2-methyl-1-propanol	78-83-1	C4H10O	74.12	99
	3-(methylthio) propanol	505-10-2	C4H10SO	106.19	98
	3-methyl-1-pentanol	589-35-5	C6H14O	102.17	99
	2-phenylethanol	60-12-8	C8H10O	122.16	99
	Benzyl alcohol	100-51-6	C7H8O	108.14	99
	Isoamyl alcohol	123-51-3	C5H12O	88.15	98
ESTERS	Ethyl butyrate	105-54-4	C6H12O2	116.16	99
	Ethyl-2-methylbutyrate	7452-79-1	C7H14O2	130.18	98
	Ethyl-3-methylbutyrate	108-64-5	C7H14O2	130.18	98
	Ethyl hexanoate	123-66-0	C8H16O2	144.21	99
	Hexyl acetate	142-92-7	C8H16O2	144.21	98
	Ethyl lactate	97-64-3	C5H10O3	118.13	98
	Ethyl octanoate	106-32-1	C10H20O	172.26	99
	Ethyl decanoate	101-38-3	C12H24O2	200.32	99
	2-phenyl-ethyl acetate	103-45-7	C10H12O2	164.2	98
	Ethyl acetate	141-78-6	C3H8O2	88.11	99
	Ethyl miristate	124-06-1	C16H30O2	256.42	98
	Isoamyl acetate	123-92-2	C7H14 = 2	130.19	95
VOLATILE FATTY ACIDS	2-methylbutyric acid	116-53-0	C5H10O2	102.13	98
	3-methylbutyric acid	503-74-2	C5H10O2	102.13	99
	Butyric acid	107-92-6	C4H8O2	88.11	99
	Hexanoic acid	142-62-1	C6H12O2	116.16	99
	Octanoic acid	124-07-2	C8H16O2	144.21	98
	Decanoic acid	334-48-5	C10H20O2	172.26	98
	Dodecanoic acid	143-07-7	C12H24O2	200.32	99

(continued)

Table 1
(continued)

Chemical group	Compound	CAS No.	Formula	Molecular weight *M*/(g/mol)	Purity (> %)
VOLATILE PHENOLS	Guaiacol	90-05-1	C7H8O2	124.14	98
	4-ethylphenol	123-07-9	C8H10O	122.16	99
	4-vinylphenol	2628-17-3	C8H8O3	120	nd
	4-vinylguaiacol	7786-61-0	C9H10O2	150.17	98
	Vanillin	121-33-5	C8H8O3	152.15	99
	Eugenol	97-53-0	C10H12O2	164.2	99
C6-COMPOUNDS	1-hexanol	111-27-3	C6H14O	102.17	98
	E-2-hexen-1-ol	928-95-0	C6H12O	100.16	96
	Z-2-hexen-1-ol	928-94-9	C6H12O	100.16	90
	E-3-hexen-1-ol	928-97-2	C6H12O	100.16	97
	Z-3-hexen-1-ol	928-96-1	C6H12O	100.16	97
INTERNAL STANDARD	4-nonanol	5932-79-6	C9H20O	144.25	95

- SPE columns Chromabond HR-P; volume 6 mL; content of sorbent 500 mg (polystyrene-divinylbenzene copolymer, 50–100 μm); material PP with PE filter elements from Machery-Nagel, Düren, Germany.
- Enzymes with glycosidase activities (β-glucosidase, α-arabinosidase, α-rhamnosidase, β-apiosidase): Rapidase® Revelation Aroma (Erbslöh, Germany) or AR 2000 (DSM food specialties, Seclin, France) can be used to released aglycons.
- Internal Standard: 4-nonanol at 300 μg/mL in absolute ethanol.

2.2 Solutions

Citrate–phosphate buffer (0.1 mol/L; pH = 5.0)

1. Prepare 800 mL of distilled water in a suitable container.
2. Add 29.41 g of Sodium Phosphate Dibasic to the solution.
3. Add 20.09 g of Citric Acid monohydrate to the solution.
4. Adjust the solution to a final desired pH using Citric acid or Sodium Phosphate.
5. Add distilled water until the volume is 1 L.
6. Store at 4 °C to 6 °C.

Pentane-Dichloromethane azeotrope (2:1)

1. Prepare 500 mL of Pentane (99.9%).
2. Add 250 mL of Dichloromethane (99.8%).
3. The mixture must be distilled at 29 °C, under atmospheric pressure.
4. Store at 4 °C to 6 °C.

3 Methods

3.1 Sample Preparation

Sample preparation and extraction method of the free and glycosidically bound aroma compounds are performed according to the methods described in Oliveira et al. [3, 15]. A total of 500 g of grapes are pressed by Termomix^{R} (Speed 4 during 15 s).

1. Must sample (200 mL) should be centrifuged ($RCF = 9660$, 20 min, 4 °C).
2. Filter the supernatant through a glass wool bed.
3. It is possible to freeze the sample (−20 °C) until analysis.

3.2 Columns Conditioning

Solid-phase extraction is performed using SPE columns Chromabond.

1. The resin of the column should be previously washed with 10 mL of dichloromethane before use (flow rate of about 1.5 mL/min, *see* **Note 1**); after, allow the cartridges to air dry during 5 min.
2. Then, the resin needs to be pre-conditioned with 5 mL of methanol and 10 mL of aqueous alcoholic solution (10%, *v*/*v*). The sample must be applied quickly before the resin dries (*see* **Note 1**).

3.3 Extraction of Free Fraction

1. A total of 75 mL of must should be passed through the resin (*see* **Note 1**) for free and bound fractions extraction after the addition of 10 μL (300 μg/mL) of 4-nonanol as internal standard solution to the must sample.
2. Wash the resin with 10 mL of ultrapure water.
3. Allow the cartridges to air dry (about 5 min) before eluting the free and bound fractions.
4. Elute the free fraction with 5 mL of pentane–dichloromethane azeotrope (*see* **Note 1**).
5. The pentane–dichloromethane extract should be dried over anhydrous sodium sulfate (*see* **Note 2**) or by freezing (*see* **Note 3**) and concentrated to 200 μL by solvent evaporation with a nitrogen stream prior to analysis.

3.4 Extraction of Glycosidically Bound Fraction

1. After the elution of the free fraction, add to the cartridge 7 mL of ethyl acetate to elute the glycosidically bound fraction (*see* **Note 1**).
2. Concentrate the ethyl acetate eluate to dryness in a Multivapor™ from Buchi (40 °C) with agitation using ALAMO tubes (*see* **Note 4**).
3. The extract needs to be resuspended in 100 μL of phosphate-citrate buffer solution and the residual volatile compounds of the free fraction should be eliminated by successive washing (5×) with about 100 μL of azeotrope. Finally, the solvent residues should be completely eliminated by evaporation under nitrogen stream.
4. Add to the extract 200 μL of a solution of 70 mg/mL ofRapidase® Revelation Aroma (Erbslöh, Germany) or AR 2000 (DSM food specialties, Seclin, France) in 0.1 mol/L citrate–phosphate buffer (pH = 5.0).
5. Incubate the mixture in a water bath at 40 °C, for 16 h (overnight) in hermetically sealed ALAMO glass tubes. Then, place the tubes on ice.
6. Released aglycons should be extracted by successive (5×) liquid–liquid extraction (LLE) with pentane-dichloromethane azeotrope, after addition of 10 μL (300 μg/mL) of 4-nonanol as internal standard solution.
7. Dry the organic phase (free compounds) over anhydrous sodium sulphate and concentrated to 200 μL with a nitrogen stream prior to analysis.

Figure 1 shows the experimental design used to develop the SPE sample preparation.

3.5 Gas Chromatography-Mass Spectrometry

Gas chromatographic analysis of volatile compounds can be performed using an Agilent GC 6890 N gas chromatograph coupled to an Agilent 7000 C triple quadrupole mass spectrometer as follows:

1. Inject 1 μL of the concentrated extract into a capillary column, coated with DB-Wax Ultra Inert (50 m × 0.25 mm i. d., 0.2 μm film thickness, Agilent) or a similar stationary phase.
2. The temperature of the injector (split/spitless with EPC control) should be 250 °C.
3. Hold the oven temperature at 60 °C, for 2 min, then program a ramp to rise from 60 °C to 234 °C, at 3 °C/min, then hold it at 5 °C/min until 250 °C, and finally program the oven for 10 min at 250 °C.
4. The carrier gas should be helium N60 (Air Liquide) at flow rate of 1 mL/min.

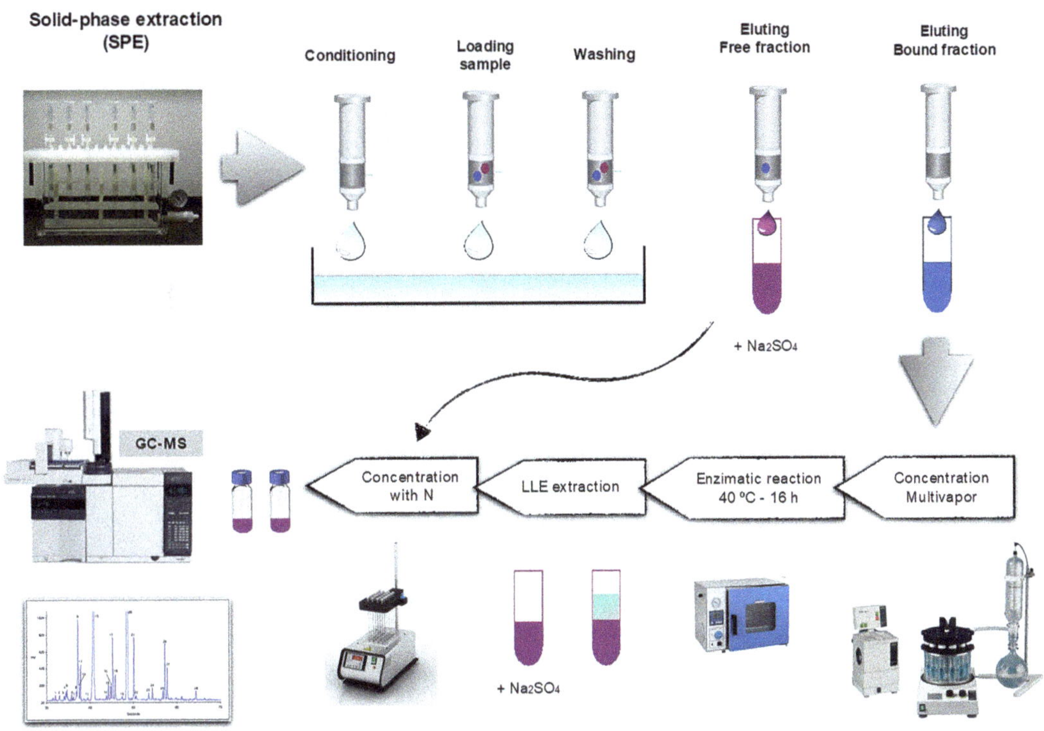

Fig. 1 Experimental design used to develop the SPE sample preparation

5. The detector should be set to electronic impact mode (70 eV), with an acquisition range from 29 m/z to 360 m/z, and an acquisition rate of 610 ms.

3.6 Compounds Identification and Quantification

Identification of volatile compounds should be performed using the GC-MS library, such as the ChemStation and Musshunter Software (Agilent), by comparing mass spectra (Wiley and NIST libraries) and retention times with those of pure standard compounds when available, or by comparing the retention indices and mass spectra with those with those reported in the literature [23, 24]. All of the compounds should be quantified as 4-nonanol equivalents. Semi-quantitative data can be obtained by calculating the relative peak area in relation to that of the internal standard (4-nonanol).

4 Notes

1. The flow rate to be applied to the SPE columns, both for conditioning, passing the must and eluting the free and glycosidically bound fractions, is about 1.5 mL/min, i.e., a drop every 3 s with a vacuum pump.

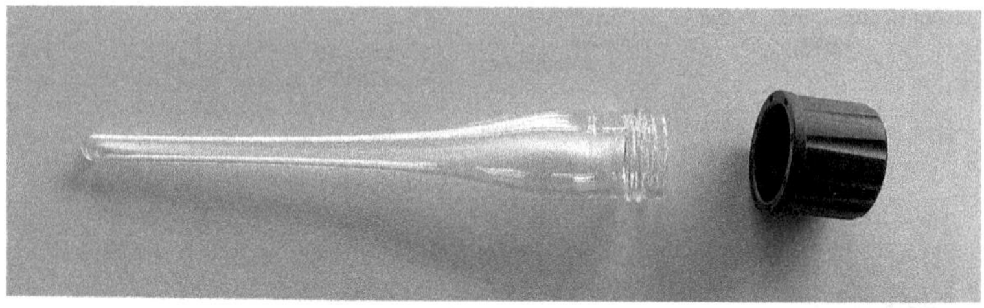

Fig. 2 New tubes for Multivapor manufactured by ALAMO (Madrid, Spain)

2. Dried over anhydrous sodium sulfate: sodium sulfate is added to the tube containing the extract to remove the water. Subsequently, the extract is transferred to another tube.
3. When the free faction is eluted, it is possible to freeze the extract to eliminate the water. Subsequently, the extract is transferred to another tube.
4. For this specific procedure, the tubes of Multivapor were designed for this study with the aim to improve the LLE extraction (enzymatic reaction in the same tube of concentration process) (Fig. 2). The tubes were manufactured by ALAMO (Madrid, Spain).

Acknowledgments

This study was supported by the Portuguese Foundation for Science and Technology (FCT) under the scope of the strategic funding of UIDB/04469/2020 unit.

References

1. Williams PJ, Strauss CR, Wilson B, Massy-Westropp RA (1982) Novel monoterpene disaccharide glycosides of *Vitis vinifera* grapes and wines. Phytochemistry 21:2013–2020
2. Günata YZ, Bayonove CL, Baumes RL, Cordonnier RE (1985) The aroma of grapes. Localisation and evolution of free and bound fractions of some grape aroma components cv. Muscat during first development and maturation. J Sci Food Agric 36:857–862
3. Oliveira JM, Oliveira P, Baumes RL, Maia MO (2008) Volatile and glycosidically bound composition of Loureiro and Alvarinho wines. Food Sci Technol Int 14:341–353
4. Sefton MA, Francis IL, Williams PJ (1993) The volatile composition of chardonnay juices: a study by flavor precursor analysis. Am J Enol Vitic 44:359–370
5. Günata YZ, Bayonove CL, Baumes RL, Cordonnier RE (1985) The aroma of grapes. I. Extraction and determination of free and glycosidically bound fractions of some aroma components. J Chromatogr 331: 83–90
6. Voirin SG, Baumes RL, Sapis JC, Bayonove CL (1992) Analytical methods for monoterpene glycosides in grape and wine. J Chromatogr A 595:269–281
7. López R, Aznar M, Cacho J, Ferreira V (2002) Determination of minor and trace volatile compounds in wine by solid-phase extraction and gas chromatography with mass spectrometric detection. J Chromatogr A 966:167–177
8. Boido E, Lloret A, Medina K, Farina L, Carrau F, Versini G, Dellacassa E (2003) Aroma composition of *Vitis Vinifera*

cv. Tannat: the typical red wine from Uruguay. J Agric Food Chem 51:5408–5413

9. Ibarz MJ, Ferreira V, Hernández-Orte P, Loscos N, Cacho J (2006) Optimization and evaluation of a procedure for the gas chromatographic–mass spectrometric analysis of the aromas generated by fast acid hydrolysis of flavour precursors extracted from grapes. J Chromatogr A 1116:217–229
10. Andujar-Ortiz I, Moreno-Arribas MV, Martín-Álvarez PJ, Pozo-Bayón MA (2009) Analytical performance of three commonly used extraction methods for the gas chromatography–mass spectrometry analysis of wine volatile compounds. J Chromatogr A 1216:7351–7357
11. Paolini M, Tonidandel L, Moser S, Larcher R (2018) Development of a fast gas chromatography–tandem mass spectrometry method for volatile aromatic compound analysis in oenological products. J Mass Spectrom 53:801–810
12. Vilanova M, Fandiño M, Frutos-Puerto S, Cancela JJ (2019) Assessment fertigation effects on chemical composition of *Vitis vinifera* L. cv Albariño. Food Chem 278:636–643
13. Panighel A, Flamini R (2015) Solid phase extraction and solid phase microextraction in grape and wine volatile compounds analysis. Sample Prep 2:55–65
14. Williams PJ, Strauss CR, Wilson B, Massy-Westropp RA (1982) Use of C18 reversed-phase liquid chromatography for the isolation of monoterpene glycosides and nor-isoprenoid precursors from grape juice and wines. J Chromatogr A 235:471–480
15. Vilanova M, Genisheva Z, Bescansa L, Masa A, Oliveira JM (2012) Changes in free and bound fractions of aroma compounds of four *Vitis vinifera* cultivars at the last ripening stages. Phytochemistry 74:196–205
16. Vilanova M, Rodríguez I, Canosa P, Otero I, Gamero E, Moreno D, Talaverano I, Valdés E (2015) Variability in chemical composition of *Vitis vinifera* cv *Mencía* from different geographic areas and vintages in Ribeira sacra (NW Spain). Food Chem 169(15):187–196
17. Dungey KA, Hayasaka Y, Wilkinson KL (2011) Quantitative analysis of glycoconjugate precursors of guaiacol in smoke-affected grapes using liquid chromatography tandem mass spectrometry based stable isotope dilution analysis. Food Chem 126:801–806
18. Allen D, Bui AD, Cain N, Rose G, Downey M (2013) Analysis of free and bound phenolics in wine and grapes by GC-MS after automated SPE. Anal Bioanal Chem 405:9869–9877
19. Schneider R, Charrier F, Moutounet M, Baumes R (2004) Rapid analysis of grape aroma glycoconjugates using Fourier-transform infrared spectrometry and chemometric techniques. Anal Chim Acta 513:91–96
20. Hampel D, Robinson AL, Johnson AJ, Ebeler SE (2014) Direct hydrolysis and analysis of glycosidically bound aroma compounds in grapes and wines: comparison of hydrolysis conditions and sample preparation methods. Aust J Grape Wine Res 20:361–377
21. Vilanova M, Oliveira JM (2012) Application of gas chromatography on the evaluation of grape and wine aroma in Atlantic viticulture (NW Iberian Peninsula). In: Salih B (ed) Gas chromatography in plant science, wine technology, toxicology and some specific applications. InTech, Rijeka, pp 109–146
22. Vilanova M, Oliveira JM, Rivas R, Alonso JC, Martínez-Zapater JM, Ibáñez J, Cacho J (2018) El Potencial Aromático de las Variedades de Vid cultivadas en Galicia. Ed: Xunta de Galicia-Consellería de Medio Rural, Santiago de Compostela
23. Di Stefano R, Borsa D, Maggiorotto G, Corino L (1995) Trepeni e polifenoli di uve aromatiche a frutto colorato prodotte in piemonte, L'Enotecnico, p 75–85
24. Di Stefano R, Bottero S, Pigella R, Borsa D, Bezzo G, Corino L (1998) Precursori d'aroma glicosilati presenti nelle uve di alcune cultivar a frutto colorato. L' Enotecnico, p 63–74

Chapter 9

Wine Volatilomics

José Sousa Câmara, Rosa Perestrelo, Cristina V. Berenguer, and Jorge A. M. Pereira

Abstract

In this chapter, we attempt to bring together in a readable and accessible form the wine volatiles patterns highlighting, in a comprehensive way, their differentiated origins. In this context, it is aimed to describe the volatile compounds i) biosynthesized in the grapes few of them are responsible for the impact aroma of some grape varieties and wines, such us monoterpenes, thiols, and pyrazines, that persist unchanged into wine; ii) the originated during the alcoholic fermentation through fermenting yeasts metabolism, enzymatic reaction and lactic acid bacteria action, and from transformation of grape-specific precursors, namely alcohols, ethyl and alkyl esters, aldehydes, and fatty acids, which contribute substantially to the flavor and to the base wine aroma, and iii) the volatiles formed during the wine storage and/or aging through several chemical reactions, including redox phenomena, Maillard reactions, and Strecker degradation, abiotic transformation of precursor compounds in wine, microbial spoilage, and difusion from oak, from which we can highlight the furanic compounds, lactones and acetals (dioxanes and dioxolanes), among others. The most important and dominant volatile compounds derived from these three main pathways are discussed in some detail.

Key words Wine, Volatilomics, Varietal volatiles, Fermentative volatiles, Aging volatiles

1 Introduction

Wine is one of the most popular, appreciated, and consumed beverages in the world. Since from a few years ago Italy, France, and Spain are the top wine-producing countries worldwide, while the United States, France, and Italy were the leading consumers (almost 82 million hectoliters). In terms of per capita consumption, and despite the heterogeneity of wine consumption behaviors across geographical regions, Portugal followed by Italy and France occupy the top positions in the ranking. The slight decrease registered in recent years in terms of consumption is influenced

José S. Câmara, Rosa Perestrelo, Cristina V. Berenguer, and Jorge A. M. Pereira have contributed equally to this chapter

María Ángeles Pozo-Bayón and Carolina Muñoz González (eds.), *Wine Analysis and Testing Techniques*, Methods and Protocols in Food Science, https://doi.org/10.1007/978-1-0716-3650-3_9,

by the significant increases in wine prices related to the energy crises and the war in Ukraine, in addition to the high inflation and global supply chain disruptions caused by the COVID-19 pandemic, which in turn lead to a spike in distribution and production costs.

Among the different factors that contribute to the acceptability of the wine by consumers, the flavor—determined mainly by the color, appearance, sugars, organic acids, and the tactile sensations in the mouth, and the aroma—determined by volatile compounds whose concentration in the wine have to be above the olfactory perception and recognition threshold to be perceived by the olfactory organs, constitute themselves as fundamental parameters, being directly associated with the chemistry of the entire winemaking process. Its acceptability is also influenced by its composition in health benefits nutrients. In addition to the polyphenolic compounds which related with antioxidant, anti-inflammatory, anti-apoptotic, and anti-proliferative properties, among others (Fig. 1), with potential benefits in the prevention of various non-communicable diseases such as cancer, cardiovascular, and neurodegenerative diseases (Alzheimer's and Parkinson's), vitamins (namely C and E), nitrogen compounds (amino acids, biogenic amines, proteins), carbohydrates, fatty acids (linoleic, linolenic, palmitic), organic acids (tartaric, malic), minerals (Na, K, P, and Ca), among others, constitute the main primary and secondary metabolites related to the nutraceutical and functional value of wines [1].

Most of these metabolites can also be found in beer, coffee, bread, vegetables, cheese, spices, and other foodstuffs. Water and ethanol are the major wine components, accounting for ≈ 97% (w/w) (Fig. 2). The compounds responsible for the wine color

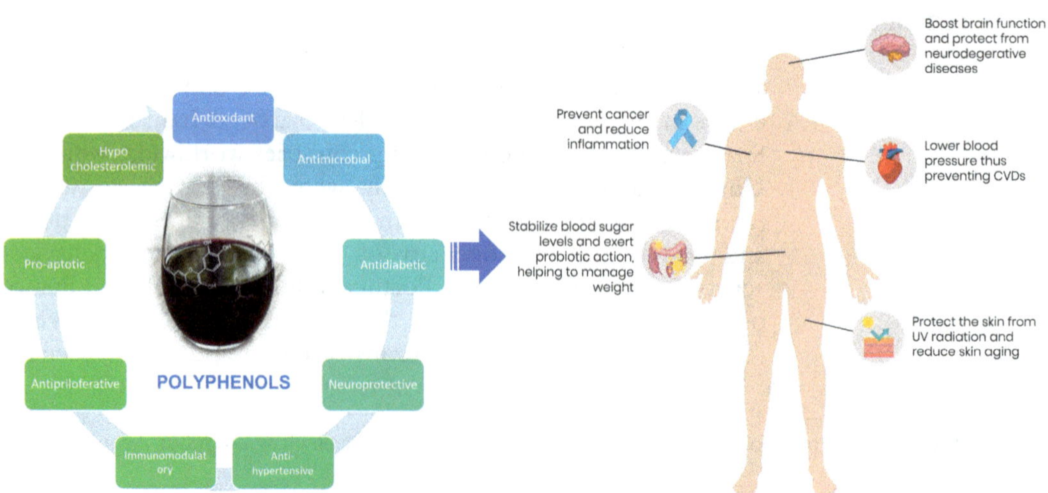

Fig. 1 Potential biological effects and health benefits of polyphenols

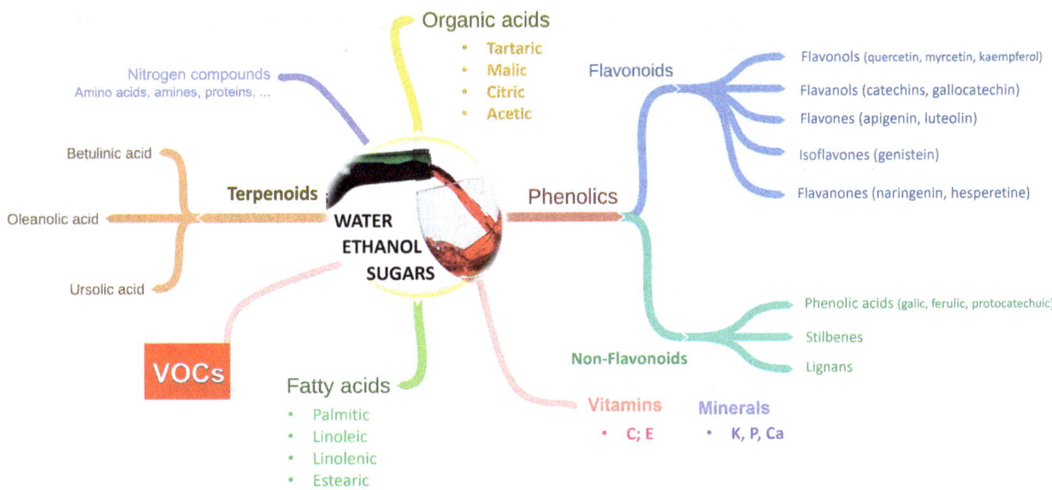

Fig. 2 Main primary and secondary metabolites related to functional and nutraceutical values of grapes and wines Potential health benefits of polyphenols

and flavor are typically present at concentrations lower than 10 g/L, while the wine aroma compounds are found at part-per-trillion (ng/L) levels. Carbohydrates, namely the hexoses fructose (a ketose) and glucose (an aldose) are the primary substrates for yeasts during alcoholic fermentation. Despite its conversion to ethanol and carbon dioxide, some residual sugar will still be detectable in some wines due mainly to incomplete fermentation, back-sweetening of wine with sucrose or grape must, hydrolysis of glycosides during storage, and extraction from oak. In wine, the sugars can contribute to perceived sweetness [2].

Fatty acids are non-volatile weak organic acids, except for acetic acid, formed by microbial metabolism or biosynthesized in the grapes. In wines, tartaric and malic acids are generally found at high concentrations (2–7 g/L), followed by citric, succinic, and lactic acids. In wine, fatty acids determine the wine pH, which affects the microbial stability, appearance, and chemical stability of wine, and in addition, they have direct impacts on taste [3].

Ammonium ($NH4^+$), the main source of usable nitrogen during alcoholic fermentation, amino acids, oligopeptides and proteins, are the major nitrogenous compounds found in wines. Most of the amino acids in grapes and wines are primary α-amino acids and proteinogenic since they are used in the synthesis of proteins during transcription. Proline followed by arginine, valine, and alanine are the most predominant amino acids in musts and wines [4]. The sources of minerals in wines include vineyard soil, fining agents, winery equipment, and winemaking treatments. The dominant minerals found in wines are Na, K, Mg, and Ca, which accumulate during berry ripening, in addition to Fe and Cu. K and Ca may also be intentionally added in the winery during deacidification treatments. These elements are associated with redox phenomena mainly due to their catalytic action. These metals promote the formation of ROS through the activation of oxygen [5, 6]. Ascorbic acid, niacin, Vitamin B6, riboflavin, thiamoin, folate, and vitamin A are present in grapes and wines. Most of the grape vitamins are used by yeast returning to the wine at similar levels that found in grapes [7]. In the following sections, the volatile compounds biosynthesized in the grapes, produced during fermentation and formed during wine storage and aging, are highlighted.

2 Volatilomic Pattern from Varietal Origin

The varietal aroma of wines are composed of volatile secondary metabolites belonging to different chemical families, namely terpenoids, pyrazines, C_{13}-norisoprenoids, and thiols [8–10]. These compounds originate in the vine and result from grape metabolism, which is dependent on the grape variety [9].

The contribution of varietal compounds to the aroma of wine depends on their concentration and odor threshold (OT), the release of aromatic compounds from its non-odorous precursors, and the chemical modifications of volatile compounds during maturation [8, 11]. The volatiles are found in wines in a wide range of concentrations, ranging from mg/L to μg/L or even ng/L [12]. Owing to their interaction with other molecules, these compounds may contribute to the overall wine aroma even when present at concentrations below their odor thresholds (OTs) [13]. In addition, if the concentration of a target volatile is too low, its contribution to the aroma is small; however, if the concentration is too high, the aroma of other components can be masked, which is detrimental to the overall wine aroma [14]. The viticultural practices and the vineyards' edaphic and climatic conditions also influence the wine aroma [8, 11].

2.1 Origin

Volatile compounds in grapes can be found in free and glycosylated forms. The glycosylated form is the most predominant and consists of aromatic precursors and odorless non-volatile compounds. At harvest time, most wine grapes have no smell, which makes odorless precursors the most abundant compounds in the grapes. Non-volatile precursors consist of organic molecules of glycosidic, amino acid, and lipidic nature [8] and play an important role in the characteristics of many wines [9]. These molecules provide volatile compounds through hydrolytic processes that occur during winemaking, in which the bond between the volatile compound and sugar moiety breaks and releases the free volatile compounds into the wine (primary aromas). Different cultivars exhibited significant qualitative and quantitative differences in volatile composition. The planting environment, related to the climate, soil, and light conditions, as well as the grape variety and yeast species, influences greatly the nature and concentration of volatile compounds by controlling grape growth and fermentation processes. Accordingly, these factors influences the wine aroma [10]. During fermentation, enzymes, yeasts, and lactic acid bacteria can also hydrolyze the glycosidic bonds of aromatic precursors being released to the wine. In addition, ethyl esters of fatty acids, acetate esters, fatty acids and in lower extent ketones and aldehydes are also biosynthesized (secondary aromas). Similarly, during this stage, the amino acids can originate higher alcohols, via the Ehrlich pathway (transamination, decarboxylation, and reduction by alchool desidrogenase), which are generaly associated with “*solvent*” and “*fusel*” odor descriptors. The aromatic complexity of wines is boosted during wine aging in oak wood barrels, resulting in aromatic compounds originating from wood (tertiary aromas) [8, 12].

2.2 Chemical Classes of Varietal Compounds

2.2.1 Terpenoids

In most of wines monoterpenoids (C_{10} compounds), sesquiterpenoids (C_{15} compounds), and C_{13}-norisoprenoids are important contributors for wine aroma [9, 10]. These compounds are typically categorized as *terpenoids*, and their molecular formula conforms to the general formula $(C_5H_8)_n$. Such as carotenoids and sesquiterpenes, share their biosynthesis starting from farnesyl pyrophosphate [8]. Furthermore, the biosynthesis of mono- and sesquiterpenes is based on the formation of isoprene C_5 units, dimethyl allyl diphosphate (DMAPP), and isopentenyl diphosphate (IPP) [9]. In wines, terpenoids undergoes chemical reactions and rearrangements given rise other terpenes, which explains the aging aroma of certain wines [8, 15]. The chemical structures of some common terpenoids found in wines are represented in Fig. 3.

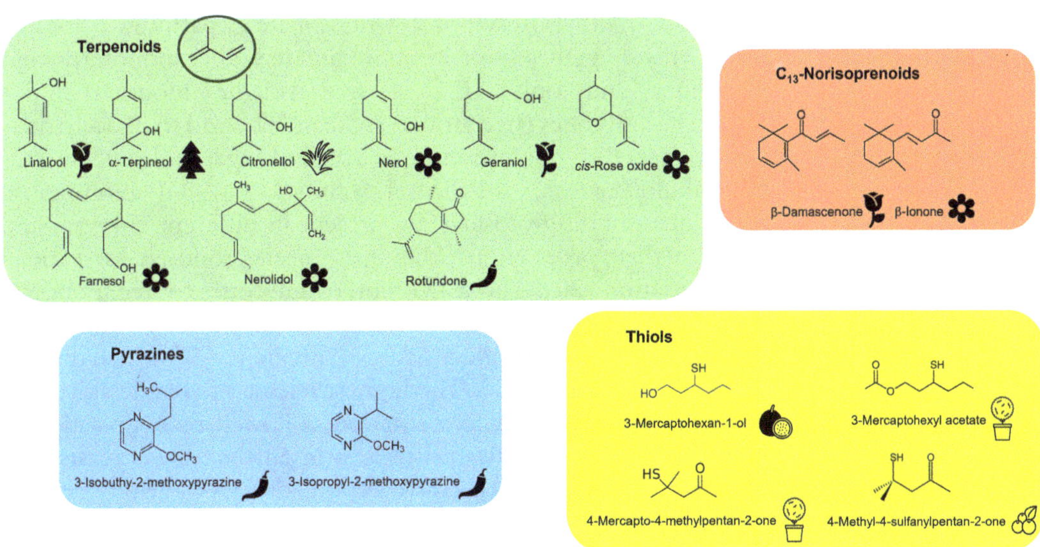

Fig. 3 Chemical groups from varietal origin and chemical structures of some varietal compounds

Monoterpenoids (C_{10} compounds) are the most abundant among terpenoids. They contribute to the varietal aroma profiles of wines with floral, fruity, and citrus notes [9, 10]. These compounds can be found in most grape cultivars, mainly in grape skin, and are synthesized during grape ripening. Their concentration and composition are affected by fruit maturity [16]. Monoterpenoids are sensitive to viticultural practices, soil type, light exposure, UV-B radiation, water deficit, basal leaf removal, crop thinning, and pests [17]. Higher temperatures and adequate sunlight are beneficial for the accumulation of monoterpenes in grapes [16]. Generally, the most important monoterpenes in wines are linalool, citronellol, geraniol, nerol, and α-terpineol [8, 10].

Based on the concentration of monoterpenes, grape varieties can be divided into muscat, non-muscat, and non-aromatic [10]. These compounds have a low OT perception; however, in grapes and wines of the Muscat family, they significantly exceed their OT and present concentrations greater than 4 mg/L, constituting themselves as aromatic impact compounds. The highest levels of monoterpenoids are found in Muscat of Alexandria, Muscat de Frontignan, Muscat Ottonel, and Muscat Blanc, among which the highest concentration can be as high as 6 mg/L [10]. The concentration of monoterpenes in non-muscat grapes ranges between 1 and 4 mg/L, whereas in non-aromatic grapes, it is less than 1 mg/L [18]. Monoterpenoids are also important odorants in other white varieties, such as Albariño, Gewürztraminer, Chardonnay, Müller-Thurgau, Riesling, Scheurebe, Sylvaner, and Traminer. In red wines its contribution to flavour is not significant [8, 9].

In addition to free and volatile monoterpenoids, grapes also contain non-volatile monoterpenoid precursors in glycosylated forms [8, 9]. Terpene glycosides were the first glycosidic compounds identified in grapes [19] and were synthesized through the 2-C-methyl-d-erythritol-4-phosphate pathway (MEP) [10]. Muscat grapes contain approximately 90% monoterpenoid glycosides and only 10% of free volatile monoterpenoids [19]. The aglycones (free aroma) are mainly bound to disaccharides that connect β-d-glucopyranose to a second sugar molecule, such as α-l-arabinofuranose, α-l-rhamnopyranose, or β-d-apiofuranose [9]. Since not all glycosides are present in different cultivars and differ in their proportions, they have been proposed for varietal differentiation [20]. The conversion of aglycones to free monoterpenoids can be carried out via acidic or enzymatic hydrolysis by enzymes, particularly β-glucosidases from grapes [9], which directly release free volatiles [21]. The highest release of volatile aglycones occurs during alcoholic fermentation, followed by malolactic fermentation, by *Saccharomyces* yeasts, non-*Saccharomyces* yeasts, and lactic acid bacteria [8, 9]. However, the kinetics of acid hydrolysis have been reported to be too low for most of the released monoterpenoids [9, 22].

Sesquiterpenes (C_{15} compounds) are present in grapes and wine. Farnesol and nerolidol are the most abundant compounds being associated to floral aromas. Under acidic conditions, the contents of farnesol and nerolidol decrease during aging, which may contribute to the balsamic and spicy aroma of Corvina grapes and wines [23]. Rotundone, biosynthesized during ripening [24], is a powerful odorant with a spicy aroma (black or white pepper). It was found in red varieties, such as the Australian cool-climate Shiraz grapes, Cabernet Sauvignon, Graciano, and Pinot Noir, among others.

2.2.2 C-13 Norisoprenoids

C_{13}-Norisoprenoids are come from the oxidative degradation of carotenoids, which is thought to be influenced by an increase in light penetration through agronomic practices [15, 25, 26]. The accumulation of C_{13}-norisoprenoids seems to occur earlier than optimal technological ripening. This feature has been used to mitigate the effects of climate change on wine quality, as early-harvested grapes produce wines with lower alcohol content and optimum aromatic quality [8, 27].

The metabolites and degradation products of carotenoids formed during berry ripening may undergo further enzymatic transformation via glycosyltransferases. This results in an increase in non-volatile C_{13}-norisoprenoid glycoconjugates and sugar production in berries [15, 28]. Moreover, hydrolysis of bound C_{13}-norisoprenoids, such as β-damascenone, often produces odorless ketones and polyols, which require a further reaction to produce the final free volatile compound [21, 29]. In grapes, the non-volatile precursors of C_{13}-norisoprenoids are generally present at concentrations significantly higher than those of the free compounds [30]. Conversion to the final aroma-active compound is thought to occur via acid catalysis during winemaking and bottle aging [31].

The aroma of C_{13}-norisoprenoids ranges from floral and fruity to petrol and fly spray [15], according to its concentration. The most important compounds are those with a megastigmane structure, which possesses a wide range of odors, from floral and fruity to scents reminiscent of tobacco and kerosene. The most abundant in grapes and wines are those with pleasant aromas, particularly β-damascenone and β-ionone, both with very low OTs [32]. However, depending on their concentration and OT perception, some compounds can negatively contribute to the aromatic profiles of wines [12]. Other important C_{13}-norisoprenoids include 3-oxo-α-ionol (tobacco aroma), β-damascone (tobacco and fruity), and 3-hydroxy-β-damascone (tea and tobacco), but their levels in wines are reported to be very low, and their OTs have not yet been defined [8, 23].

β-Damascenone has a rose, cooked apple, or honey aroma and can be detected in grapes as free volatile at a concentration up to 9 μg/kg [33]. This compound has been found to act as an aroma enhancer for ethyl esters associated with the berry fruit aroma. Moreover, it can mask herbaceous pyrazine-related characteristics [34]. β-Damascenone is produced via oxidative cleavage of the carotenoid neoxanthin, followed by enzymatic transformation, and glycosylation of the intermediates grasshopper ketone and megastigma-6,7-dien-3,5,9-triol, and lastly acid-catalyzed conversion to the aroma compound [35]. It can react with free sulfur dioxide (SO_2) at wine pH to give an odorless sulfonic acid derivative. Consequently, this results in the loss of 50% of β-damascenone at room temperature over 30 days in the presence of 80 mg/L SO_2 [36]. β-Ionone is a metabolite of β-carotene found in its free form in grapes [37]. Its aroma threshold is similar to that of β-damascenone, and it has a violet raspberry and rose aroma. Moreover, this compound seems to be more important for the aroma of red wine than that of white wine [15]. The chemical structure of C_{13}-norisoprenoids found in wines are represented in Fig. 3.

2.2.3 Thiols

Volatile thiols represent a large family of compounds that are of great importance for the organoleptic quality of wines. Owing to their low OTs perception, these compounds have strong effects on the sensorial properties of wines, despite their very low concentrations [8, 9, 38]. Thiols seem to be dependent on vine agronomic practices [39]. Moreover, they are chemically unstable, easily oxidized, and can be transformed into undesirable aromas even in reducing environments [8, 40]. Despite this, a lack of oxygen can reduce odor generation [41], and as such, the storage and aging conditions are determinants of the thiol concentration in wines [9].

Varietal thiols have been identified in a wide range of grape varieties [9]. In grapes, occurs only as non-odorous precursors of cysteine (Cys) and glutathione (GSH), which can be detected in the pulp and peel [8]. Thiol precursors are produced in vine plants as a detoxification mechanism through the conjugation of unsaturated alkenals, forming 3-sulfanylhexan-1-ol precursors and alkenones, forming 4-methyl-4-sulfanylpentan-2-one precursors with GSH. The tripeptide GSH is then hydrolyzed to dipeptide cysteine-glycine (Cys-Gly) and Cys. Thus, GSH, Cys-Gly, and Cys are precursors of 3-mercaptohexanol and 4-methyl-4-sulfanylpentan-2-one in grapes. The acetylated form of 3-sulfanylhexan-1-ol, 3-sulfanylhexyl acetate, is formed by the acetylation of 3-sulfanylhexan-1-ol after this compound is produced during fermentation [9, 42]. Free thiols are mainly released during alcoholic fermentation if the yeast possesses a specific lyase activity for the S-cysteine and glutathione conjugates and is absent in must [8]. Yeasts can take up thiol precursors from grape juice and cleave the conjugated precursor, releasing the corresponding free thiols, using ammonium as a nitrogen source and pyruvate [9, 43].

Thiols can be classified as highly volatile and low-volatility compounds. Highly volatile compounds such as carbon sulfide, ethanethiol, methanethiol, and hydrogen sulfide, are associated with aroma defects [9]. Low-volatility compounds are the most desirable thiol compounds that contribute to the enhancement of the sensorial quality of wines [44, 45]. This group includes compounds with high molecular weights and low volatility, which are found at very low concentrations but above the threshold value in wine, such as 4-methyl-4-sulfanylpentan-2-one, 3-sulfanylhexan-1-ol, and its derivative, 3-sulfanylhexyl acetate. These compounds are among the most important thiols associated with the aroma of white wines [46] and have been detected in many varieties, such as Sauvignon Blanc, Macabeo, Gewürztraminer, Riesling, Verdejo, Merlot, and Cabernet Sauvignon, in which 3-sulfanylhexan-1-ol and 3-sulfanylhexyl acetate are more prevalent than 4-methyl-4-sulfanylpentan-2-one [9, 41, 45]. The typical aroma of some of these wines with tropical notes reminiscent of grapefruit, boxwood, and passion fruit, is attributed to these powerful odorants [8]. Other varietal thiols include 4-mercapto-4-methyl-pentan-2-ol, 3-mercaptopentan-1-ol, and 3-mercaptoheptan-1-ol [9, 44].

The thiols concentration in wines depends on several factors. One of the most important factors is the concentration of thiol precursors in grapes, which is found in the skin and pulp at μg/L levels. Moreover, the concentration of these compounds depends on several factors, such as harvesting mode, SO_2 treatment, ripeness, vine nitrogen conditions, water deficit, grape variety, temperature, and grape skin tannins, among others [41, 47–50]. Additionally, the yeast strain used for fermentation is one of the most important factors affecting thiol production [51] (reviewed in [9].

Besides thiol concentration in wine, its perception is associated with the chemical composition of the wine matrix [52]. Decreased levels of esters or higher alcohols can reduce the masking effects of these compounds on minor compounds such as thiols. *Metschnikowia pulcherrima*, in combination with *Saccharomyces cerevisiae* (*S. cerevisiae*), can increase the 4-methyl-4-sulfanylpentan-2-one concentration and also reduce higher alcohol production, thereby increasing the fruitiness of wines [9, 53]. Figure 3 shows the chemical structure of some important thiols found in wines.

2.2.4 Pyrazines

Pyrazines are a class of nitrogen-containing heterocyclic compounds mainly formed by the catabolism of some amino acids such as leucine, isoleucine, and valine, and are found in grapes as free volatiles [8, 10]. They are highly odorous compounds with a very low OT, which allows them to have a significant impact on wine aroma [54]. These metabolites possess green and/or herbaceous notes at high levels that can be detrimental to high-end wines, particularly >15 ng/L in white wines and > 25 ng/L in red wines [10]. This characteristic aroma can be found in the grape varieties *Vitis vinifera* cv., Cabernet Sauvignon, Sauvignon Blanc, Cabernet Franc, and Merlot noir [55, 56]. However, the presence of this aroma is only acceptable in a few white wines and is unfavorable in some red wines, especially those from the Cabernet family [57]. Hence, it is important to understand the factors that influence the production and accumulation of pyrazines in wine grapes and eventually in the produced wine [55–60].

The biochemical processes involved in the formation of pyrazines in grapes and their degradation during grape development are affected by viticultural conditions. For instance, the levels of 3-isobutyl-2-methoxypyrazine decrease as grapes mature [59]. The concentration of pyrazines is affected by light exposure, climate, and terroir, as their concentrations increase in cooler environments and are reduced by strong light [54, 55, 59, 61]. Nevertheless, the biosynthesis and accumulation of these compounds require further exploration [55, 61].

3-Isobutyl-2-methoxypyrazine, 3-isopropyl-2-methoxypyrazine, and 3-*sec*-butyl-2-methoxypyrazine are the most commonly identified pyrazines in wines [55]. 3-Isobutyl-2-methoxypyrazine is the most abundant and the most likely to affect wine aroma, as it is often found in concentrations above its sensory threshold (2 ng/L in water) [10, 62]. In wines from cool climates, it exceeds its low OT, especially if the wines come from early harvests since it diminishes considerably with maturation [55]. The vine water status can also affect the concentrations of 3-Isobutyl-2-methoxypyrazine, as irrigation of the vines increases its content [8, 63]. The bell pepper aroma of Sauvignon Blanc and Cabernet Sauvignon wines has been correlated with the levels of this compound [64]. Furthermore, its sensory thresholds differ between white, red, and synthetic wines. The values found in red wines, for instance, ranged between 1 to 6 ng/L and 10 to 16 ng/L [55, 58]. 3-isopropyl-2-methoxypyrazine is the second-heaviest methoxypyrazine in wines with potato or earthy aroma [65]. It is found in grapes and wines, but can also be produced by microorganisms [55]. 3-*sec*-butyl-2-methoxypyrazine is associated to pea and bell pepper aroma [14]. The most common pyrazines found in wines are shown in Fig. 3 [43, 66–70] (Table 1).

Table 1
Main varietal aroma compounds detected in wines or model wines. OT: olfactory detection threshold

Chemical family	Compounds	Aroma description	OT (μg/L)	Refs.
Terpenoids				
Monoterpenes	Geraniol	Geranium, rose	30	[137]
	Linalool	Coriander seed, rose	15	[137]
	Nerol	Floral, orange flowers	15	[66]
	Citronellol	Rose, lemongrass	100	[66]
	α-Terpineol	Floral, lilac, pine	250	[67]
	cis-Rose oxide	Floral, green	8	[70]
Sesquiterpenes	Farnesol	Floral	1000	[68]
	Nerolidol	Floral	15	[137]
	Rotundone	Black or white pepper	0.016	[69]
C_{13}-Norisoprenoids				
	β-Damascenone	Apple sauce, rose	0.05	[137]
	β-Ionone	Violet	0.09	[67]
Thiols				
	3-Mercaptohexan-1-ol	Passion fruit, grapefruit	0.06	[70]
	3-Mercaptohexyl acetate	Boxwood, passion fruit	0.004	[70]
	4-Mercapto-4-methylpentan-2-one	Boxwood, broom	0.0008	[70]
	4-methyl-4-sulfanylpentan-2-one	Boxwood, blackcurrant	0.003	[43]
Pyrazines				
	3-Isobuthy-2-methoxypyrazine	Green pepper, pea pod	0.001	[70]
	3-Isopropyl-2-methoxypyrazine	Green pepper, pea pod, potato, earthy	0.002	[70]

3 Volatilomic Pattern of Fermentation

The process of wine fermentation involves the conversion of grape juice into wine through a complex chemical reaction with yeast and sugars. This process is influenced by factors such as initial sugar content and fermentation temperature, and it continues even after fermentation is completed. During this process, many volatile compounds that impact wine aroma are formed or converted from precursors previously existent in grapes, or consumed [71–73].

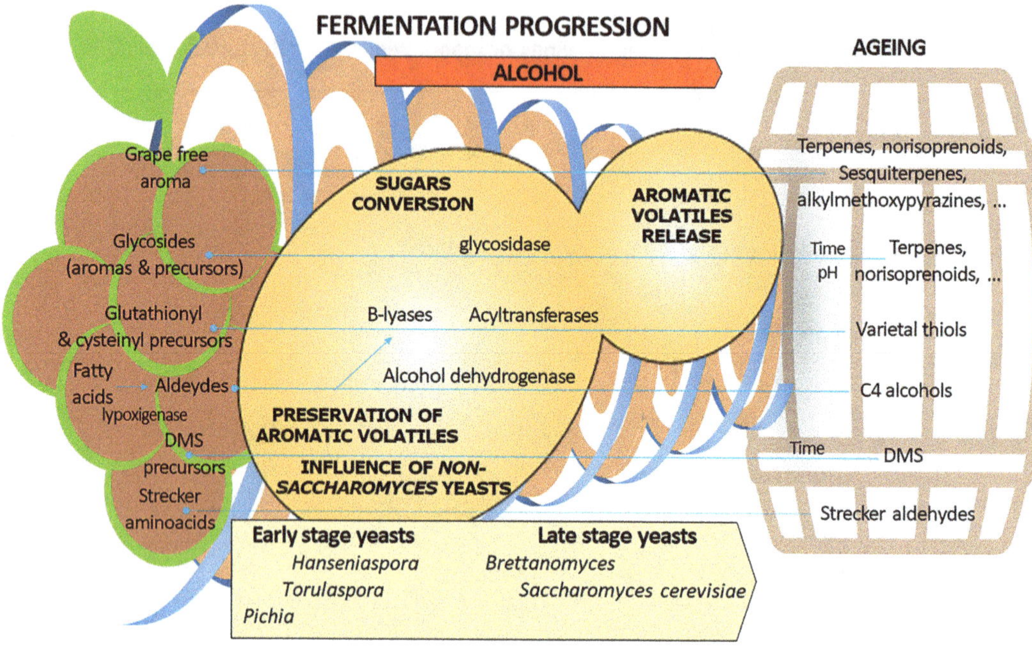

Fig. 4 overview of the different aspects influencing the final wine aroma

Therefore, fermentation plays a crucial role in determining the quality and character of wine [73, 74]. This contribution occurs at different levels and steps in the process and is affected by the quality and properties of the grapes, the fermentative microorganisms present, and the fermentation conditions [72, 75]. Controlling fermentation is essential for ensuring consistent wine quality, and mathematical models are often used to predict and optimize fermentation behavior. Fermentation temperature and initial sugar content also play a significant role in preserving volatile aromatics and fruity characters in the wine. The choice of yeast strains and environmental conditions, especially temperature, also influence the formation of substances that contribute to the taste, character, bouquet, and stability of the wine. By considering these aspects, winemakers can improve fermentation speed, shorten fermentation time, and enhance overall wine quality. In summary, the contribution of fermentation to wine flavors can be categorized as the conversion of sugars into alcohol, the release of volatile aroma compounds, the preservation of volatile aromatics, and the influence of non-*Saccharomyces* yeast strains [9]. An overview of fermentation contribution to the final wine aroma is presented in Fig. 4.

3.1 Conversion of Sugars During Fermentation

During the process of fermentation, yeasts convert the natural sugars found in grapes into ethanol, carbon dioxide, and other minor flavor-active compounds [8, 76]. This conversion is made possible by the enzymes produced by the yeast. However, the activity of these enzymes can be influenced by various factors, such as environmental and biotic stressors that may be present during fermentation [8]. To adapt to these conditions and maintain their metabolic activity, yeast cells must convert sugars into alcohol, even under challenging circumstances. However, not all yeast strains have the same capacity for sugar conversion [77]. Therefore, it is necessary to extensively screen different yeast strains to select those that have positive attributes, such as enhanced production of glycerol and esters [77]. On the other hand, it is important to avoid strains that have negative impacts, such as the overproduction of acetic acid or hydrogen sulfide [77]. In addition to *S. cerevisiae*, which is the most commonly used yeast strain in winemaking, other non-*Saccharomyces* yeasts, including *Torulaspora delbrueckii*, *Hanseniaspora* spp., *Lachancea thermotolerans*, *Metschnikowia pulcherrima*, *Pichia* spp., and *Candida zemplinina*, also play a role in sugar conversion [78]. As we will see in more detail in the next sections, these non-*Saccharomyces* yeast strains have different metabolic activities depending on the fermentation conditions, which can influence the conversion of sugars during fermentation and thus impact the diversity and complexity of wine production [77].

3.2 Release of Volatile Aroma Compounds

The choice of yeast species and strains, as well as fermentation conditions, can significantly affect the release and development of aroma compounds in wines. Grapes contain glycosylated precursors of volatile aroma compounds, such as glycosides, glutathionyl, and cysteinyl conjugates, and other non-volatile molecules. Enzymatic activities, including glycosidases produced by yeasts during fermentation, can release these bound aroma compounds, including esters, thiols, terpenes, norisoprenoids, and phenols [8, 73]. Up to 27 relevant wine aroma compounds can be considered to proceed from grape-specific precursors. Some of these compounds are immediately formed during fermentation, while others require a long aging time to accumulate. The volatile compounds released contribute to the fruity, floral, and spicy aromas of the wine [8, 73, 79]. In particular, yeasts can enhance the concentration of volatile thiols such as 4-mercapto-4-methylpentan-2-one, 3-mercaptohexan-1-ol, and 3-mercaptohexyl acetate in wines [71]. These thiols impart desirable passionfruit, grapefruit, citrus, and other aroma characters, which are especially important in Sauvignon Blanc wines but can also be relevant in other varieties such as Riesling, Semillon, Merlot, and Cabernet Sauvignon [71]. These thiols are initially present in grapes as nonvolatile forms conjugated to cysteine. During fermentation, yeasts like *S. cerevisiae* and *S. bayanus* produce a cysteine lyase enzyme that deconjugates these thiols into their volatile form. The ability to release these volatile thiols varies with the specific *Saccharomyces* strain used. Furthermore, yeasts play a crucial role in the biochemical transformation of flavor-inactive grape juice constituents into flavor-active components. One well-studied reaction is the liberation of terpenes. Monoterpene alcohols such as citronellol, geraniol, linalool, and nerol naturally occur in grapes, particularly in Muscat, Riesling, and other white varieties, contributing to fruity, estery, spicy, and vegetative aromas. However, a significant portion of these grape terpenes is covalently linked to glucose or disaccharides of glucose and other sugars, which have no flavor impact. Yeast glycosidases break down these sugar conjugates, releasing volatile terpenes and significantly impacting the wine's character. The production of glycosidases varies with yeast species and strain, with non-*Saccharomyces* yeasts like *Hanseniaspora*, *Debaryomyces*, and *Dekkera* being stronger producers of these enzymes compared to *S. cerevisiae*. Therefore, these non-*Saccharomyces* yeasts likely have a greater role in the release of terpene aromas in wines (reviewed in [71]). Additionally, as referred to in the previous section, volatile aglycones are mainly released during the fermentation steps and involve both *Saccharomyces* and non-*Saccharomyces* yeasts, as well as lactic acid bacteria [8, 9, 22].

3.3 Preservation of Volatile Aromatics

The fermentation process is crucial for preserving volatile aromatics in wine and enhancing the aromatic compounds derived from grapes [9]. Factors such as temperature control and the use of specific yeast strains influence the preservation of these volatile aromatics [8]. Proper management of fermentation conditions ensures that desired aromatic compounds are not lost or degraded, resulting in a more pronounced and complex flavor profile in the final wine [8]. This control of fermentation is vital for consistent wine quality and plays a significant role in preserving volatile aromatics and fruity characters. Additionally, the proper management of fermentation conditions, including temperature control and the choice of yeast strains, helps retain and enhance aromatic compounds from grapes, contributing to the wine's aromatic complexity and flavor profile [8, 80]. Furthermore, wine evolution continues after fermentation until the production of the final product. Controlling the fermentation process, maintaining the desired temperature, and using suitable yeast strains are crucial aspects of wine fermentation. Current understanding of yeast's actions in winemaking extends beyond the simple metabolism of grape juice sugars and includes the metabolism of grape juice sugar and nitrogen components, enzymatic hydrolysis of grape components affecting wine aroma, flavor, color and clarity, autolysis, and bioadsorption [71].

3.4 Influence of Non-Saccharomyces *Yeast Strains*

Yeasts play a fundamental role in alcoholic fermentation, converting sugars to ethanol and CO_2 and producing volatile organic compounds. *S. cerevisiae* has been extensively studied [74], but non-*Saccharomyces* strains have been found to increase the sensory complexity and flavor diversity of wines [78]. In fact, over

40 different yeast species have been found to participate in wine fermentation [76]. These strains produce different enzymatic activities and metabolic by-products compared to the dominant yeast, *S. cerevisiae*. As a result, their participation in fermentation can result in wines with more pronounced fruit aromas and a greater variety of flavors. Accordingly, the diversity of yeast strains contributes to the wine's unique character and flavor profile. For this reason, non-*Saccharomyces* strains are of commercial and oenological interest due to their positive impact on their unique abilities and sensory qualities. Studies have revealed the existence of non-*Saccharomyces* model organisms such as *Torulaspora delbrueckii*, *Hanseniaspora* spp., *Lachanthua thermotolerances*, *Mechinikovia pulcherrima*, *Candida zemplinina*, and *Pichia* spp., in combination with *S. cerevisiae* (Table 2) [74, 78]. As can be observed, different types of yeast can have different effects, both positively on winemaking and as potential sources of wine failure. Regarding this, yeasts belonging to the *Brettanomyces/Dekkera* genus are often associated with wine spoilage all over the world [81].

Yeast settlers can colonize diverse habitats, including vineyards and wineries, and form microbial communities associated with specific niches. Understanding the composition, dynamics, and maintenance of these yeast communities provides insight into the impact of yeast on grape health, grape quality, and wine sensory profile [76]. The natural microbiota associated with wine fermentation is very complex and may include other eukaryotic and prokaryotic organisms (e.g., lactic acid bacteria [8, 9]). Its composition depends on many factors, such as rainfall, grape variety, temperature, soil, viticulture methods, and irrigation [77]. Non-*Saccharomyces* yeast strains, including *Hanseniaspora* and other genera, form an important part of the native grape flora and contribute to the sensory complexity of wines [80]. In particular, *Hanseniaspora vineae* (*H. vineae*) has emerged as a promising species for quality wine production. Wines made with *H. vineae* and *S. cerevisiae* exhibit greater fruit intensity and complexity than wines made with *S. cerevisiae* alone. In turn, subtle aromatic notes have been associated to other non-*Saccharomyces* species such as *Pichia*, *Metschnikowia*, and *Torulaspora* [77, 82, 83]. Table 2 briefly describes the specific effects of including non-*Saccharomyces* yeast strains in the fermentation process.

Beyond the type of yeast, the parameters of the fermentation process can be used to modulate yeast gene expression. Under permissive conditions, for instance, in which yeasts can complete the fermentation process and consume all sugars available, cells can cope with the natural environmental and biotic stressors and maintain metabolic activity [77]. However, challenging conditions (e.g., temperature and sugar availability) can be used to modulate yeast gene expression, therefore affecting the expression of key enzymes (phenotype) in the fermentation process. Such knowledge is being

Table 2
The main non-*Saccharomyces* yeast species of oenological importance and their influence on wine fermentation

Species	Oenological impact	Refs.
Torulospora delbrueckii	↑esters and thiols; ↑glycerol, diacetyl, ethyl lactate, and ethyl acetate, ↓acetic acid; ↓acetic acid production	[86–88, 98]
Prelude™ (Torulospora delbrueckii)	↑medium-chain fatty acid esters (increases flavor complexity), ↑mannoproteins, ↓Volatile acidity, ↓toxic medium-chain fatty acids (promotes malolactic fermentation)	[88]
Zymaflore® Alpha (Torulospora delbrueckii)	↑Volume and length palate, ↑aromatic diversity and intensity, ↑3SHA, ↑3SH, ↓volatile acidity, ↓acetaldehyde, acetoin, diacetyl, and H_2S	[88]
Biodiva™ (Torulospora delbrueckii)	↑Aromatic and mouthfeel complexity, ↑aromatic esters, ↑osmotic shock resistance, ↓volatile acidity	[88]
Viniferm NS TD (Torulospora delbrueckii)	↑wine complexity, ↑aromatic spectrum, ↑ß-phenyl ethanol, ↑ß-lyase activity, ↑mannoprotein	[88]
Primaflora® VB BIO (Torulospora delbrueckii)	bioprotection	[88]
Lachancea thermotolerans	↑L-lactic acid, glycerol, and 2-phenyl-ethanol (used to acidify grape juices); ↓volatile acidity in wine	[89, 91, 94, 108, 114]
Metschnikowia pulcherrima	↑esters, terpenes, thiols, and aromatic complexity; ↑β-glucosidase activity, ↑compounds influencing the organoleptic quality of wine (e.g. medium-chain fatty acids, higher alcohols, esters—mainly ethyl octanoate, terpenols, and glycerol), ↓volatile acidity	[53, 113]
Schizosaccharomyces pombe	↑deacidification by L-malic acid degradation;	[90, 92]
Candida zemplinina	↑glycerol and succinic acid; ↑linalool, citronellol, nerolidol, geraniol, and terpenes (wines with fructophilic and complex aromas); ↓acetic acid and higher alcohols	[77, 104]
Candida stellata	↑glycerol content; ↑wine aroma profile; ↓ acetaldehyde, acetoin, glucose and fructose	[95, 109]
Candida cantarellii	↑glycerol content	[110]
Candida pulcherrima	↑wine aroma profile	[99]
Hanseniaspora spp.	↑acetate esters and terpenes; biogenic amine adsorption	[77]
Hanseniaspora vineae	↑ acetate esters (e.g., 2-phenylethyl acetate) and phenylpropanoids (e.g., 2-phenylethyl and benzyl alcohols); ↑fruit intensity aromas described as banana, pear, apple, citric fruits, and guava; ↓branched-chain higher alcohols, fatty acids, and ethyl esters	[97, 102]

(continued)

Table 2 (continued)

Species	Oenological impact	Refs.
Hanseniaspora guilliermondii and *Hanseniaspora uvarum*	↑1-propanol, 2-phenylethyl acetate and 3-(methylthio)propionic acid; ↓ethyl hexanoate, pentanoic acid, free fatty acids, 2-methyltetrahydrothiophen-3-one and acetic acid-3-(methylthio)propyl ester	[106]
Hansenula anomala	↑higher alcohols and acetate and ethyl esters; ↓C6 alcohols	[77]
Hansenula uvarum (*K. apiculata*)	Simulation of natural fermentation (improvement of aroma complexity, ↓ethyl acetate esters)	[103, 105]
Pichia guillermondii	↑color stability and 4-ethyl-phenol production	[77]
Pichia kluyveri	↑varietal thiols and esters (↑glycerol and ethanol yield)	[77, 112]
Zygosaccharomyces bailii	↑polysaccharides; acetic acid	[77]
Kluyveromyces thermotolerans	↓acetic acid production; Enhancement of titratable acidity	[100]
Issatchenkia orientalis	↓malic acid content	[101]
Pichia fermentans	↑and more complex aroma	[93]
Pichia kluyveri	↑varietal thiols (e.g., 3-mercaptohexyl acetate, 2-phenylethyl acetate, and ethyl octanoate)	[77, 85]
Debaryomyces vanriji	↑geraniol concentration	[96]
Schizosaccharomyces spp. + *Saccharomycodes* spp. + *Pichia* spp.	Influence on sensorial and physico-chemical properties of wines	[107]
Mixed 'wild' yeasts	↑and more complex aroma	[72, 111]

3SH 3-sulfanylhexan-1-ol, *3SHA* 3-sulfanylhexylacetate

unveiled with the final aim of obtaining wines with adjusted aromatic profiles [53, 72, 77, 84–114].

3.5 Overview of Key Aromatic Compounds During Fermentation

Benzenoids, especially phenylpropanoids, are present in grapes in very low concentrations as free volatiles and can be found in important levels in the form of glycosylated precursors. Glycosylated forms of volatile phenols such as eugenol, guaiacol, vanillin, and their derivatives, can contribute to the spicy aroma of wines, without being subjected to aging in contact with oak wood. Moreover, the balsamic aroma associated with the terroir of many wines can be attributed to these glycosylated compounds. Other precursors of volatile phenols include coumaric and ferulic cinnamic acids, which are mainly found in grape skins. However, these compounds require enzymes with decarboxylation activity, which can be inhibited in the presence of catechins. Therefore, the presence of these volatile phenols is more important in wines of short maceration, such as white wines, than in those of long maceration, as in the case of red wines [8, 70].

C_6-compounds are free volatiles directly derived from grapes and have herbaceous, leafy, cut-grass aromas [8, 9]. These compounds consist of alcohols and aldehydes, such as 1-hexanol, (Z)-3-hexenol, (E)-3-hexenol, and (E)-2-hexenal, which can be found at concentrations above the corresponding sensory thresholds [42]. Due to their high OT, alcohols are among the least odorant volatiles [115]. C_6-compounds are considered varietal aromas since they come from the grape variety [116]. Although, these compounds are common in all grapes. In addition, they can be considered pre-fermentative aromas, as their content increases from the moment in which the berry is separated from the cluster and in all the other pre-fermentative stages. In grapes, C_6-compounds can be found as non-odorous glycosylated precursors, but their concentration is much lower than those of the free forms. Moreover, oxygen triggers their formation and activates enzymatic systems, generating aldehydes and alcohols from polyunsaturated fatty acids, which are favored by the temperature [8]. The roles of these compounds in the final perception of wine aroma depend on their concentrations, which consequently depend on the grape variety and other climatic and viticultural factors [9]. For instance, the concentration of 1-hexanol can range from 1320 to 13.800 μg/L, with a sensory threshold of 8000 μg/L. In turn, cis-3-hexenol ranges from 8 to 711 μg/L, with a sensory threshold of 400 μg/L [67]. At low concentrations, these compounds can contribute to the complexity and typicity of some wines. Nonetheless, at high concentrations, they show undesirable effects, such as imparting undesirable green aromas and as a depreciator of clear, fruity notes in both white and red wines [9].

4 Volatilomic Pattern of Storage and Aging Processes

The volatile composition of wines is significantly affected by the aging process; specifically, red wines such as Madeira wines, Porto wines, and Sherry have been the subject of main research [117–120]. During the aging process, several interactions can occur and molecular can migrate from oak to wine contributing to the fortified wine's distinctive sensory qualities. The quality of the fortified wine aging bouquet can be influenced by several parameters, such as grape variety, wine-making process, and oak features (e.g., age of cask, geographical origin, species of oak, seasoning of the staves, and toasting), among others [121, 122]. Călugăr et al. [123] evaluate the influence of untoasted oak chips, toasted oak chips, and untoasted barrels on the volatile profile of Muscat Ottonel Wine during aging. The untoasted medium improved the terpenes

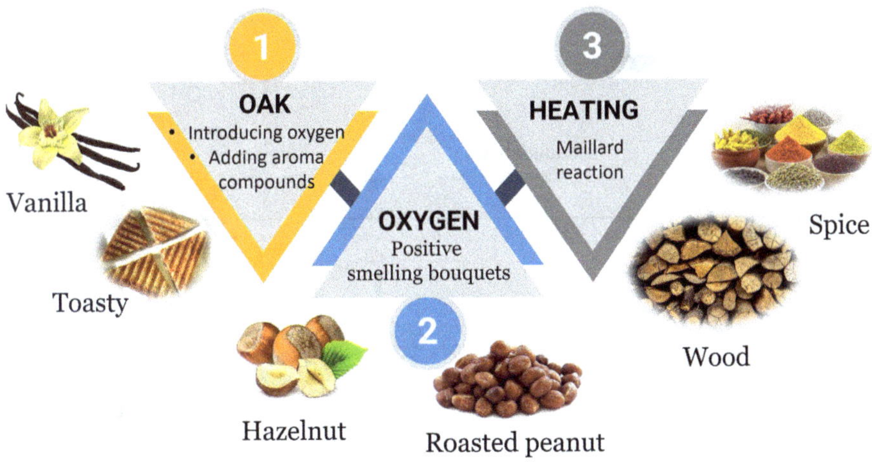

Fig. 5 Mechanisms and odor descriptors related to aged Madeira wines

and alcohols that contribute Muscat Ottonel its characteristic scent, whereas acetovanillone and p-vinyl guaiacol, two volatile wood components, were increased by lightly roasted oak chips. Moreover, as fortified wines age, the concentration of some fermentative (e.g., ethyl esters, alcohols) and varietal (e.g., terpenoids, norisoprenoids) compounds reduces, causing the wine to become less fruity and fresh. In contrast, other aromas, such as those of caramel, dried fruits, spices, toast, and wood, begin to emerge (see Fig. 5). The key mechanisms relating to these characteristics are the Maillard processes, Strecker degradation caramelization, microbial activity, and diffusion from the oak [124–126]. Port wine is often matured for a long time, up to many decades, in oak barrels. During this process, Port wines develop diverse tastes of fruit, chocolate, caramel, and spices as a result of aging [122]. In addition, the concentration of volatile organic compounds changes significantly, consequently, the wine's color and fragrance vary significantly, being more noticeable with extended aging. A similar pattern is observed in the volatile profile of Madeira wines during the aging process. The results obtained in a recent study performed by Perestrelo et al. [127] represent a suitable approach for encouraging changes in the estufagem process and evaluating the impact of storage on the volatile profile.

Furanic compounds, lactones, volatile phenols, and acetals are the most chemical families related to the aging process, being mentioned as potential Madeira wine aging markers, as their levels increase significantly with aging time and can be used to prevent fraud. Compounds, such as diethoxymethane, 1,1-diethoxyethane, 1,1-diethoxy-2-methyl-propane, 1-(1-ethoxyethoxy)-pentane, trans-dioxane, cis-dioxane, cis-dioxolane, trans-dioxolane, 2-propyl-1,3-dioxolane, 2-furfural, 5-methylfurfural, 5-hydroxymethylfurfural, cis-oak-lactone, trans-oak-lactone,

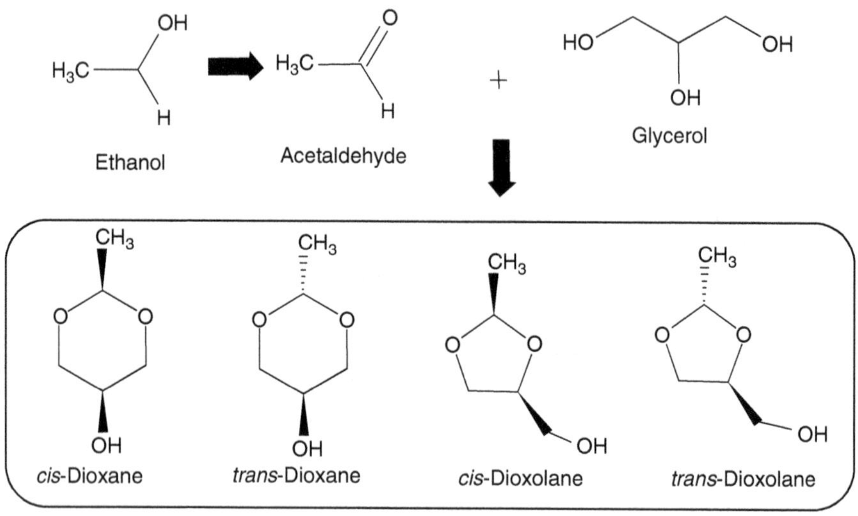

Fig. 6 Formation of heterocyclic acetals by the acetalization reaction between acetaldehyde and glycerol. (Adapted from Alti-Palacios et al. [11])

sotolon, eugenol, guaiacol, and p-ethylphenol have been reported in several studies as potential Madeira wine aging markers [120, 121, 128–130]. The four heterocyclic acetal alcohols mentioned before as potential aging markers were formed as a result of the acid-catalyzed condensation process between aldehydes and alcohols, which is preferred at higher pH levels, as exemplified in Fig. 6 [131].

From a sensorial point of view, fortified wines are substantially influenced by volatile compounds since they promote a variety of experiences through tastes and odors that encourage customer acceptance or rejection. In general, a single volatile compound at a concentration greater than its odor threshold is enough to provide a specific odor and contribute to the overall wine aroma [132]. Perhaps, furanic compounds and acetals are at higher concentrations especially in older wines, their contributions to fortified wine aroma are not expected, since these chemical families showed higher odor thresholds (mg/L) [67]. Pyrolysis of carbohydrates, Maillard reaction-induced dehydration of sugars, and caramelization are the main pathways involved in the formation of the furanic compounds [118, 133, 134]. Silva et al. [118] evaluated the effect of forced-aging on Madeira wine aroma by gas chromatography-olfactometry (GC-O) and identified a pool of Maillard by-products, like as 2-furfural, 5-methyl-2-furfural, 5-(ethoxymethyl)-2-furfural, methional, which clarifies the baked, brown sugar, and nutty aroma descriptors of Madeira wine. Moreover, lactones are crucial odorants compounds of Madeira wines aged in oak casks, mainly the oak-lactone, pantolactone, sotolon, and γ-lactones, which results from the cyclization of the corresponding hydroxycarboxylic acids [132]. Several studied conducted on older Madeira and Porto wines reported sotolon as an important key odorant due to its lower odor threshold (10 μg/L) [67, 118], and related with their distinctive caramel, curry and nutty odor. Câmara et al. [135] verified that the concentration of sotolon in sweet Madeira wines (Malvasia) is higher compared to dry Madeira wines (Sercial), suggesting that sotolon concentration is influenced by sugar level. Moreover, in this study, it was also verified that the concentration of sotolon is also influenced by age since their concentration increase from 100 (6-year-old) to 1000 μg/L (25-year-old). Regarding to the volatile phenols, specially 4-ethylguaiacol and 4-ethylphenol, can be linked with olfactory defects in wine if they are present at a concentration above their odor threshold, 33 and 440 μg/L, respectively [124]. However, at low concentrations, this chemical family contribute positively to young wines with a distinctive aged character by transferring scent notes of spices, smoke, and leather [136].

5 Wine Volatilomics Analysis

The wine volatilomic analysis can generally be classified into (i) TARGET focused on a specific group of intended metabolites thereby requiring subsequent quantification and identification, or (ii) UNTARGET focused on the detection/identification of a variety of metabolites that allow fingerprints or model standards without the need to quantify or identify specific metabolites (Fig. 7).

5.1 Techniques Used in Wine Volatile Extraction

Wine is a complex mixture of aromatic compounds that can contain several hundred different volatile organic compounds (VOCs), ranging from a few ng L^{-1} to g/L. Therefore, the extraction and comprehensive analysis of wine VOCs are challenging. Moreover, for many, if not all, wine producers, knowledge of the consistency

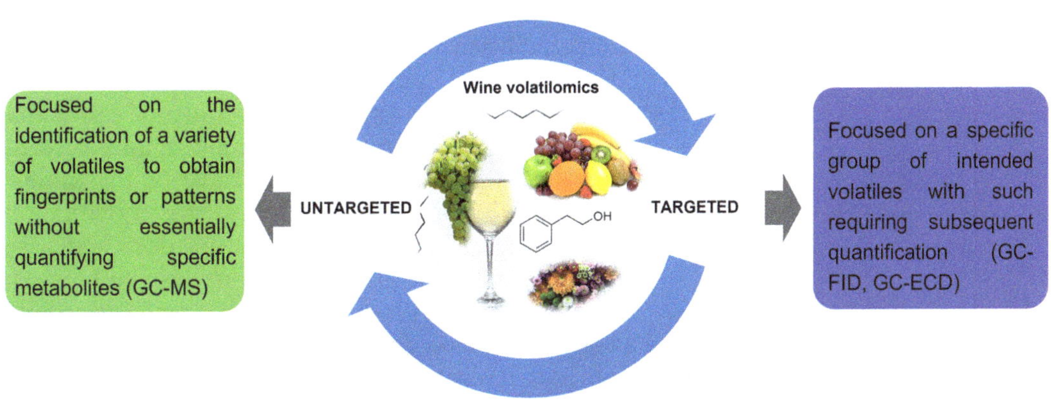

Fig. 7 Modes of volatilomics analysis

of wines produced year after year is very important. This requires the use of robust methodologies that can deliver consistent results only reflecting changes in the samples analyzed and not variations due to the experimental layout or methodological parameters. Accordingly, only a few techniques for extracting volatile organic compounds (VOCs) from wine have been reported in the literature. Table 3 lists the commonly used methods, including LLE, SPE, HS-SPME, HSSE, SBSE, and MASE, which have been used since 2018.

LLE

The use of LLE to extract wine VOCs involves a large sample volume (5–8 mL) mixed with an organic solvent, DCM [137, 138] or diethyl ether [139], using a magnetic stirrer or ultrasonic bath (US). US is more efficient than magnetic stirring, reducing the extraction time from 15 to 5 min. A subsequent centrifugation step separates the organic phase containing the wine VOCs, which are further dried with anhydrous Na_2SO_4 and analyzed by GC-MS. A reference internal standard, often an external alcohol, is often added to the wine sample prior to extraction.

SPME

SPME is one of the most successful extraction procedures used for VOCs analysis. The reasons for this are certainly related to the fact that it is a solventless procedure with a simple and straightforward experimental layout that generates few residues. On some occasions, the sample is not disturbed during extraction. In the headspace (HS) mode, which is the most commonly used mode, the fiber is only exposed to the HS of the sample, which is usually agitated upon pH and ionic strength adjustment. In the case of wine, and according to studies reported in the literature in the last 5 years, this HS mode is preferred, as is the triple DVB/CAR/PDMS fiber. This type of fiber has been demonstrated to be the most beneficial for analyzing wine VOCs, offering advantages such as higher extraction efficiency, improved aroma profiling, ability to capture a wide range of compounds, and enhanced detection of VOCs [140–142]. Overall, the remaining experimental procedure involved some optimization according to the wine sample. Accordingly, in the selected reports presented in Table 3, the sample volume ranged from 0.5 [143] to 12 mL [144], the equilibration time is usually fast (10–15 min), and the extraction step with stirring of the sample at a controlled temperature is often between 30 min and 60 min, although longer extraction times are sometimes used [144]. Following HS-SPME extraction, the trapped VOCs were loaded and separated by GC/MS.

Table 3
Overview of main extraction approaches used for wine samples

Extraction technique	Extraction conditions	Analytes	Refs.
LLE	8 mL of wine +2.4 μg 4-nonanol (IS) + 400 μL DCM. Extraction by stirring with a magnetic stir bar (15 min, room temperature). Cooling (10 min, 0 °C). Centrifugation (5 min, RCF = 5118, 4 °C). Organic phase recovered. Drying with anhydrous Na_2SO_4. Aromatic extracted recovered (200 μL). GC-MS analysis	VOCs	[137]
LLE	5 mL wine after alcoholic fermentation +100 μL 4-methyl-2-pentanol (IS) + 1 mL diethyl ether (solvent) placed in an ultrasonic bath (5 min). Centrifugation (3 min, 4000 rpm). Na_2SO_4 was added to remove any water from the non-polar layer. Centrifugation (3 min, 4000 rpm). GC-FID analysis: 3 μL extracted sample	VOCs	[139]
LLE	8 mL wine +3 μg 4-nonanol (IS) + 400 mL of DCM. Extraction by stirring with a magnetic stir bar (15 min, room temperature). Cooling (10 min, 0 °C). Centrifugation (5118 × *g*, 5 min, 4 °C) for organic phase removal. Aromatic extract recovered (200 μg L^{-1}) and dried with anhydrous Na_2SO_4. GC-MS analysis	VOCs	[138]
HS-SPME	0.5 mL wine aliquot +10 μL of IS mixture (d_4–3-methyl-1-butanol, d_3-hexyl acetate, d_{13}–1-hexanol, d_5-ethyl nonanoate, d_5–2-phenylethanol, d_{19}-decanoic acid) with 4.5 mL water +2 g NaCl. DVB/CAR/PDMS fiber, equilibration (10 min, 50 °C), extraction (45 min, 50 °C, 500 rpm). GC-MS analysis	VOCs	[143]
HS-SPME	12 mL wine +2 g NaCl. DVB/CAR/PDMS fiber, equilibration (15 min, 60 °C, with stirring), extraction (105 min, 60 °C, with stirring). GC-MS analysis	VOCs	[144]
HS-SPME	8.5 mL wine +50 μL IS (mixture of 17.7 mg L^{-1} methyl-2-methylbutyrate, 20 mg L^{-1} benzyl alcohol $_{13}C6$, 45 mg L^{-1} methyl octanoate, 185 mg L^{-1} heptanoic acid, 20 mg L^{-1} 3,4-dimethylphenol and 16.3 mg L^{-1} hexanal) diluted to a final volume of 25 mL with a hydroalcoholic solution (13.5% ethanol +3.5 g L^{-1} tartaric acid, pH adjusted to 3.5). Extraction:10 mL of this dilution +3.5 g NaCl	VOCs	[12]

(continued)

Table 3 (continued)

Extraction technique	Extraction conditions	Analytes	Refs.
	(DVB/CAR/PDMS fiber, 60 min, 40 °C, 500 rpm). GC-MS analysis		
HS-SPME	5 mL wine + IS (ethyl decanoate and 3-methyl-1-butanol) + NaCl (10%). DVB/CAR/PDMS fiber, extraction (105 min, 50 °C, with stirring). GC-MS analysis		[140]
HS-SPME	4 g sample (grape pomaces, grape stems, or lees) + 2 g of NaCl +5 mL of ultra-pure water +10 μL 4-methyl-2-pentanol (IS, 250 μg/L). DVB/CAR/PDMS fiber, extraction (45 min, 40 °C, 800 rpm). GC-MS analysis	VOCs	[152]
SPE	25 mL wine passed through preconditioned PP-DVB cartridges (40–120 μm, 5 mL of volume, 0.2 g of adsorbent phase) + 4-nonanol (IS). Column rinsed with 25 mL water to eliminate sugars, acids, and other polar compounds. Free fraction eluted with 15 mL pentane:DCM (2:1 v/v). Extracts concentrated by distillation in a Vigreux column and nitrogen stream to 100 μL (kept at −20 °C until GC–MS analysis)	VOCs	[145]
SPE	50 mL wine and grapes +10 μL 2-octanol (IS, 420 mg/L in EtOH) diluted with 50 mL deionized water. Solution loaded on an SPE cartridge (1 g sorbent previously activated with 20 mL MeOH and equilibrated with 20 mL water). Cartridge washed with 15 mL water. Free volatile compounds eluted with 10 mL DCM and concentrated under nitrogen stream to 200 μL before GC injection. Bound compounds eluted with 20 mL MeOH. Solvent evaporation under vacuum to dryness. Bound compounds dissolved in 5 mL citrate buffer (pH 5) with 100 mg of PVPP + 200 μL enzyme preparation AR2000 (70 mg/mL in citrate buffer)	Terpenoids, norisoprenoids, benzenoids	[153]
SPE	50 mL grape and wine +10 μL 2-octanol (IS, 420 mg L^{-1} in EtOH) diluted with 50 mL deionized water. Solution loaded on an SPE cartridge (1 g sorbent previously activated with 20 mL MeOH and equilibrated with 20 mL water). Cartridge washed with 15 mL water. Free volatile compounds eluted with 10 mL DCM. Bound compounds eluted by 10 mL MeOH. Free volatile compounds concentrated	VOCs	[154]

(continued)

Table 3 (continued)

Extraction technique	Extraction conditions	Analytes	Refs.
	under a gentle nitrogen stream to 200 μL and ready for injection. Methanolic-phase eluted with 10 mL DCM, dried with Na_2SO_4 (1 spatula) and concentrated to 200 μL under a gentle nitrogen steam		
LLE + SPE	LLE: aliquot of wine sample extraction with DCM three times (v/v, 5%, 15 min). Organic phases pooled and kept at −20 °C until required. Ag^+SPE: MetaSep IC-Ag cartridge conditioned with 10 mL of DCM. LLE extracts loaded onto the conditioned cartridge. Cartridge washed with 10 mL DCM, 20 mL ACN, and 10 mL DCM. Elution with 5 mL water, 20 mL aq. l-cysteine solution (10 g/L), and 10 mL DCM. Elutes (H_2O, l-cysteine solution, and DCM) collected and stirred (15 min, 900 rpm). Organic phase pooled, dried over anhydrous Na_2SO_4, and concentrated under nitrogen for further GC-MS analysis	Volatile thiols	[155]
SBSE	5 mL wine vinegar +1.67 g NaCl +10 μL 4-methyl-2-pentanol (IS, 1045 mg/L). Extraction with a 10-mm long stir bar coated with a 0.5-mm PDMS layer (60 min, 62 °C). After 5 min at room temperature, the stir bar was removed with tweezers, rinsed with water, and dried with lint-free tissue paper. The stir bar was transferred into a glass tube (60 mm long, 6 mm o.d. and 4 mm i.d.) and placed in the autosampler tray for thermal desorption and GC–MS analysis	VOCs	[149]
SBSE	SBSE: 8 mL centrifuged must sample (15 min, 4500 rpm) + 1 mL standards solution +25 μL 2-octanol solution (5 μL 2-octanol/100 mL EtOH) + 2.5 g NaCl. HS extraction with a PDMS-coated stir bar (6 h, 60 °C, 500 rpm) + borosilicate magnetic stirrer. Stir bar removed from the sample, rinsed with distilled water, and dried with tissue paper. Stir bar transferred into a thermal desorption tube for GC-MS analysis ^{m}SBSE: 0.8 mL centrifuged must sample (15 min, 4500 rpm) + 0.1 mL of standards solution +25 μL 2-octanol solution (5 μL 2-octanol/100 mL EtOH) + 8.1 mL water. Extraction with PDMS coated stir bar (0.5 mm	VOCs SBSE provided a higher extraction of volatile compounds than ^{m}SBSE	[148]

(continued)

Table 3 (continued)

Extraction technique	Extraction conditions	Analytes	Refs.
	film thickness, 10 mm length) + ethylene glycol coated stir bar (10 mm length, 32 μL phase volume) (6 h, 60 °C, 500 rpm). Stirs bars removed, rinsed with distilled water, and dried with cellulose tissue. Stir bars transferred into a thermal desorption tube for GC-MS analysis		
MASE	15 mL wine in a glass crimp-cap, headspace vial with a magnetic stir bar (3 × 8 mm) + EtOH solution containing labelled compounds as IS (50 μL, equivalent to ~17 μg/L of each d_6–1,8-cineole, d_6-linalool, d_6-α-terpineol, d_2-β-citronellol, d_7-geraniol, d_4-β-damascenone, d_3-α-ionone, and d_3-β-ionone) + MASE apparatus (membrane bag, sealing ring, and cone insert). Crimp capped and placed on a cooler tray held at 10 ° C. Before analysis, the MPS Robotic Pro added hexane/acetone (2:1, 0.9 mL) into the membrane bag, the vial was stirred (45 min, 35 °C), cooled (15 min, 10 °C), and 2 μL extract injected for GC-MS analysis	Monoterpenes and C_{13}-norisoprenoids	[151]

ACN acetonitrile, *DCM* dichloromethane, *EtOH* ethanol, *GC-FID gas chromatography* with flame-ionization detection, *GC-MS* gas chromatography-mass spectrometry, *HS* headspace, *LLE* liquid-liquid extraction, *MASE* membrane-assisted solvent extraction, *MeOH* methanol, *^{m}SBSE* multi-stir bar sorptive extraction, *Na_2SO_4* sodium sulphate, *NaCl* sodium chloride, *PP-DVB* polypropylene-divinylbenzene, *PVPP* polyvinylpolypyrrolidone, *SBSE* stir bar sorptive extraction, *SPE* solid-phase extraction, *SPME* solid-phase microextraction

SPE

In SPE, very large volumes of wine (25–50 mL) are loaded through a preconditioned cartridge containing the sorbent particles that will retain the wine VOCs. Subsequently, a washing step is used to discard interferents, such as sugars, acids, and other polar compounds [145], and a final elution step allows the recovery of the wine VOCs. Often, methanol is used as elution solvent to facilitate the extract concentration under nitrogen stream before injection in the GC-MS.

SBSE

In SBSE, the sorbent is incorporated in the magnetic stir bar and extraction occurs while the solution is being stirring under a controlled temperature. Often, a layer of PDMS is used to coat the stir bar, enabling a large area of contact with the sample. After the extraction occurs, the stir bar is just removed, dried with a lint-free tissue paper and the trapped analytes eluted by placing the stir bar in a suitable solvent and stir. The eluted analytes can then be analyzed or concentrated if necessary [146, 147]. An alternative route was used by [148, 149] for wine VOCs analysis, by transferring the stir bar for a desorption tube and desorb the trapped VOCs directly to the GC-MS [148, 149], therefore skipping the methanol elution.

MASE

Membrane-assisted solvent extraction is a small-scale liquid-liquid extraction using a polymer membrane (low-density polyethylene) as a small reservoir of the organic phase for the target analytes enrichment [150]. This technique is quite simple, enabling a high enrichment factor and its combination with a large volume injection GC-ECD, further extends the analytical performance of the methodology. The application of MASE to wine samples characterization was elegantly shown by Pisaniello et al. [151] that used a robotic system to control the MASE extraction. However, it should be mentioned that MASE is more suitable for targeted volatomics, and the authors were specifically interested in the extraction of wine monoterpenes and C13-norisoprenoids [12, 137–140, 143–145, 148, 149, 151–155].

5.2 Instrumental Analysis

Several instrumental techniques, including those based in (i) gas chromatography (GC-FID, GC-MS, and GCxGC-ToFMS) for the qualitative and quantitative evaluation of the chemical composition of the volatiles; (ii) gas chromatography-tandem olfactometric detection which allows the individual aromatic characterization of the wine volatiles; and (iii) electronic nose very useful in chemometric modeling of the global volatile composition, are commonly used in the analysis—detection, separation, identification, and quantification—of wine volatile compounds. From those, chromatographic techniques are the golden standard analytical instruments to this end. Chromatographic analysis has been extensively used to investigate wine's chemical properties, as well as to monitor winemaking steps, to identify and quantify volatile compounds (e.g., terpenoids, norisoprenoids, esters, alcohols), non-volatile compounds (e.g., phenolic compounds, flavonoids, anthocyanins), organic acids, sugars, organic metal-containing compounds, pesticides, among others target analytes [156–160]. The main chromatographic analysis used in wine's investigation requires a sample preparation step for the isolation and/or preconcentration of target analytes before separation/detection analytical techniques. Gas chromatography (GC), liquid chromatography (LC), and multidimensional techniques (e.g., GC × GC; LC × LC) are the most used analytical approaches in the investigation of wine's properties and taste. In this sense, GC-MS has been highly used in the investigation of the changes in volatile composition during the aging process with the purpose of identifying potential key aromas and aging markers [126, 129, 161]; to evaluate the impact of several winemaking techniques on the volatile composition as well as on the sensory properties [162], to explore the effect of different fermenting yeast strains on the aroma components [163], to quantify odor-active carbonyls (aldehydes, Strecker aldehydes, unsaturated aldehydes, ketones) in fortified wines [164], among other examples. GC-ECD has been used to determine haloanisoles in wines [165], whereas GC-FID has been applied to monitor the wine quality [166], to determine free fatty acids [167] and methanol in wines [168]. Recently, Welke and collaborators [169] provided an easy and economical step-by-step procedure for the adaptation of a GC-FID to a lab-made olfactometer in combination with comprehensive two-dimensional gas chromatography–mass spectrometry, (GC × GC-MS) to evaluate odor-active compounds of wine. Twenty-four odor-active compounds were identified by GC-O, while GC × GC-MS signaled additional 14 odor-active compounds, which were found as coelutions. As stated in this study, GC × GC was capable to overcome numerous restrictions when separating complex samples with high resolution, sensitivity, and peak capacity through the synergistic separation of two chromatographic columns with different stationary phases [169, 170]. Figure 8 shows a schematic overview of the potential instrumental techniques used in the detection, identification, and quantification of wine volatile compounds.

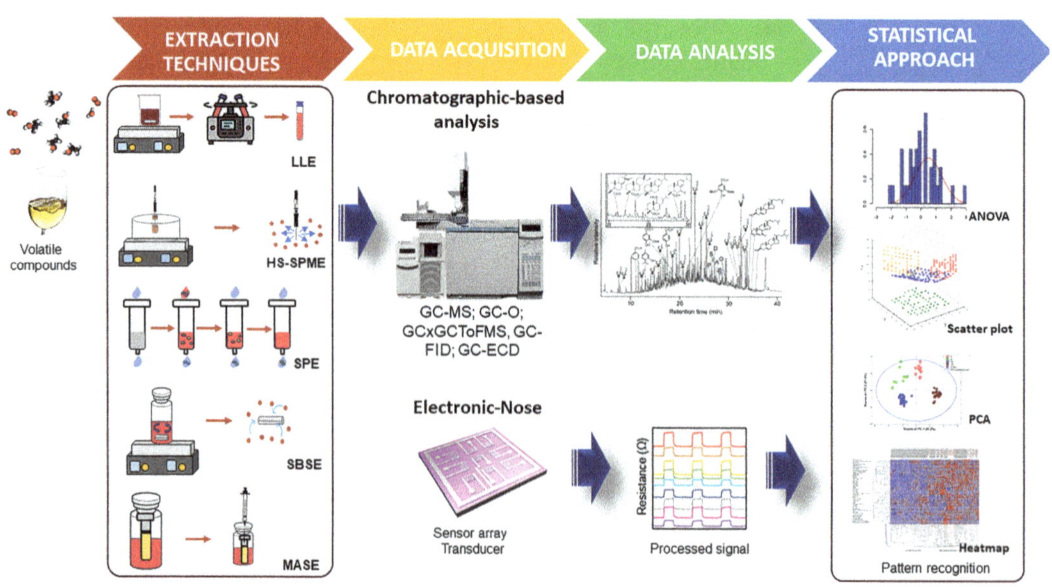

Fig. 8 Schematic workflow used in the analysis of wine volatilomics

6 Final Remarks

The wine quality is determined by two main parameters that define the characteristic properties of wines: (i) the terroir related to the grape variety/rootstock, climatic conditions (temperature, irrigation), the soil (profile, physical characteristics, water resource, fauna), to the harvest year—factors that determine the composition of the grapes and, consequently, the must and the wine and (ii) the art, the engine and the knowledge of the wine producers acquired over centuries of experience, in the vinification process used, in particular, temperature, yeast strain, must treatment, use of fermentation aids, filtration and other processes used, together with any maturation and/or aging processes. The winemaking process namely fermentation, in which the hexose sugars, glucose and fructose (fermentable sugars) present in the grape juice, are converted by yeasts into mainly ethyl alcohol and carbon dioxide, in addition to several secondary volatiles, responsible for the base wine aroma, and non-volatile metabolites, ensuring yeast growth and maintenance. The maturation of the wine, either in-cask (or vat/tanks), or in-bottle, after vinification, slow chemical reactions including slight oxidation of some existing components by the dissolved oxygen present and absorbed by wine during storage in addition to the chemical/physical extraction from the wood of the barrels, can occur influenced by the time of conservation, temperature, and humidity. All of these processes explain part of the secret behind the wine flavor and quality.

Funding

This research was supported by FCT-Fundação para a Ciência e a Tecnologia through the CQM Base Fund, UIDB/00674/2020, Programmatic Fund, UIDP/00674/2020, and by ARDITI-Agência Regional para o Desenvolvimento da Investigação Tecnologia e Inovação, through the project M1420-01-0145-FEDER-000005—Centro de Química da Madeira—CQM+ (Madeira 14-20 Program). The authors also acknowledge FCT and Madeira 14-2020 program to the Portuguese Mass Spectrometry Network (RNEM) through the PROEQUIPRAM program, M14-20 M1420-01-0145-FEDER-000008). ARDITI is also acknowledged for the postdoctoral fellowship granted to J.A.M.P. (Project M1420-09-5369-FSE-000001.

References

1. Câmara JS, Albuquerque BR, Aguiar J, Corrêa RC, Gonçalves JL, Granato D, Pereira JA, Barros L, Ferreira IC (2021) Food bioactive compounds and emerging techniques for their extraction: polyphenols as a case study. Foods 10:37
2. Liu SQ, Davis CR (1994) Analysis of wine carbohydrates using capillary gas liquid chromatography. Am J Enol Vitic 45:229. https://doi.org/10.5344/ajev.1994.45.2.229
3. Fowles GWA (1992) Acids in grapes and wines: a review. J Wine Res 3:25–41. https://doi.org/10.1080/09571269208717912
4. Bell S-J, Henschke PA (2005) Implications of nitrogen nutrition for grapes, fermentation and wine. Aust J Grape Wine Res 11:242–295. https://doi.org/10.1111/j.1755-0238.2005.tb00028.x
5. Danilewicz J (2011) Review of oxidative processes in wine and value of reduction potentials in enology. Am J Enol Vitic 63:1–10. https://doi.org/10.5344/ajev.2011.11046
6. Sirén H, Sirén K, Sirén J (2015) Evaluation of organic and inorganic compounds levels of red wines processed from Pinot Noir grapes. Anal Chem Res 3:26–36. https://doi.org/10.1016/j.ancr.2014.10.002
7. Hagen KM, Keller M, Edwards CG (2008) Survey of biotin, pantothenic acid, and assimilable nitrogen in winegrapes from the Pacific Northwest. Am J Enol Vitic 4:432–436
8. Rubio-Bretón P, Salinas MR, Nevares I, Pérez-Álvarez EP, Álamo-Sanza M, Marín-San Román S, Alonso GL, Garde-Cerdán T (2020) Recent advances in the study of grape and wine volatile composition: varietal, fermentative and ageing aroma compounds. Food Aroma Evol 22:439–463
9. Ruiz J, Kiene F, Belda I, Fracassetti D, Marquina D, Navascués E, Calderón F, Benito A, Rauhut D, Santos A et al (2019) Effects on varietal aromas during wine making: a review of the impact of varietal aromas on the flavor of wine. Appl Microbiol Biotechnol 103:7425–7450. https://doi.org/10.1007/s00253-019-10008-9
10. He Y, Wang X, Li P, Lv Y, Nan H, Wen L, Wang Z (2023) Research progress of wine aroma components: a critical review. Food Chem 402:134491. https://doi.org/10.1016/j.foodchem.2022.134491
11. Alti-Palacios L, Martínez J, Teixeira JAC, Câmara JS, Perestrelo R (2023) Influence of cold pre-fermentation maceration on the volatilomic pattern and aroma of white wines. Foods 12. https://doi.org/10.3390/foods12061135
12. del Barrio Galán R, Bueno-Herrera M, de la Cuesta PL, Pérez-Magariño S (2022) Volatile composition of Spanish red wines: effect of origin and aging time. Eur Food Res Technol 248:1903–1916. https://doi.org/10.1007/s00217-022-04014-x
13. Perestrelo R, Silva C, Camara JS (2019) Madeira Wine volatile profile. A platform to establish madeira wine aroma descriptors. Molecules 24. https://doi.org/10.3390/molecules24173028
14. Zhao P, Qian Y, He F, Li H, Qian M (2017) Comparative characterization of aroma compounds in merlot wine by lichrolut-en-based aroma extract dilution analysis and odor activity value. Chemosens Percept 10:149–160. https://doi.org/10.1007/s12078-017-9236-4
15. Black CA, Parker M, Siebert TE, Capone DL, Francis IL (2015) Terpenoids and their role in wine flavour: recent advances. Aust J Grape Wine Res 21:582–600
16. Li X, Ren J-N, Fan G, Zhang L-L, Pan S-Y (2021) Advances on (+)-nootkatone microbial biosynthesis and its related enzymes. J Ind Microbiol Biotechnol 48, kuab046. https://doi.org/10.1093/jimb/kuab046
17. Robinson A, Boss P, Solomon P, Trengove R, Heymann H, Ebeler S (2014) Origins of grape and wine aroma. Part 2. Chemical and sensory analysis. Am J Enol Vitic 65:25–42. https://doi.org/10.5344/ajev.2013.13106
18. Zhu B-Q, Cai J, Wang Z-Q, Xu X-Q, Duan C-Q, Pan Q-H (2014) Identification of a plastid-localized bifunctional nerolidol/linalool synthase in relation to linalool biosynthesis in young grape berries. Int J Mol Sci 15:21992–22010
19. Park SK, Morrison JC, Adams DO, Noble AC (1991) Distribution of free and glycosidically bound monoterpenes in the skin and mesocarp of Muscat of Alexandria grapes during development. J Agric Food Chem 39:514–518. https://doi.org/10.1021/jf00003a017
20. Ghaste M, Narduzzi L, Carlin S, Vrhovsek U, Shulaev V, Mattivi F (2015) Chemical composition of volatile aroma metabolites and their glycosylated precursors that can uniquely differentiate individual grape cultivars. Food Chem 188:309–319. https://doi.org/10.1016/j.foodchem.2015.04.056

21. Anna KH, Susan EE (2015) Glycosidically bound volatile aroma compounds in grapes and wine: a review. Am J Enol Vitic 66:1. https://doi.org/10.5344/ajev.2014.14104
22. Ugliano M, Bartowsky EJ, McCarthy J, Moio L, Henschke PA (2006) Hydrolysis and transformation of grape glycosidically bound volatile compounds during fermentation with three saccharomyces yeast strains. J Agric Food Chem 54:6322–6331. https://doi.org/10.1021/jf0607718
23. Slaghenaufi D, Ugliano M (2018) Norisoprenoids, sesquiterpenes and terpenoids content of Valpolicella wines during aging: investigating aroma potential in relationship to evolution of tobacco and balsamic aroma in aged wine. Front Chem 6:1–13
24. Mattivi F, Caputi L, Carlin S, Lanza T, Minozzi M, Nanni D, Valenti L, Vrhovsek U (2011) Effective analysis of rotundone at below-threshold levels in red and white wines using solid-phase microextraction gas chromatography/tandem mass spectrometry. Rapid Commun Mass Spectrom 25:483–488. https://doi.org/10.1002/rcm.4881
25. Chen WK, Yu KJ, Liu B, Lan YB, Sun RZ, Li Q, He F, Pan QH, Duan CQ, Wang J (2017) Comparison of transcriptional expression patterns of carotenoid metabolism in 'Cabernet Sauvignon' grapes from two regions with distinct climate. J Plant Physiol 213:75–86. https://doi.org/10.1016/j.jplph.2017.03.001
26. Hernandez-Orte P, Concejero B, Astrain J, Lacau B, Cacho J, Ferreira V (2015) Influence of viticulture practices on grape aroma precursors and their relation with wine aroma. J Sci Food Agric 95:688–701. https://doi.org/10.1002/jsfa.6748
27. Asproudi A, Ferrandino A, Bonello F, Vaudano E, Pollon M, Petrozziello M (2018) Key norisoprenoid compounds in wines from early-harvested grapes in view of climate change. Food Chem 268:143–152. https://doi.org/10.1016/j.foodchem.2018.06.069
28. Baumes R, Wirth J, Bureau S, Gunata Y, Razungles A (2002) Biogeneration of C13-norisoprenoid compounds: experiments supportive for an apo-carotenoid pathway in grapevines. Anal Chim Acta 458:3–14
29. Zoecklein BW, Wolf TK, Marcy JE, Jasinski Y (1998) Effect of fruit zone leaf thinning on total glycosides and selected aglycone concentrations of Riesling (Vitis vinifera L.) grapes. Am J Enol Vitic 49:35–43
30. Gunata YZ, Bayonove CL, Baumes RL, Cordonnier RE (1985) The aroma of grapes. Localisation and evolution of free and bound fractions of some grape aroma components cv Muscat during first development and maturation. J Sci Food Agric 36:857–862
31. Robinson AL, Boss PK, Solomon PS, Trengove RD, Heymann H, Ebeler SE (2014) Origins of grape and wine aroma. Part 1. Chemical components and viticultural impacts. Am J Enol Vitic 65:1–24
32. Sáenz-Navajas M-P, Campo E, Fernández-Zurbano P, Valentin D, Ferreira V (2010) An assessment of the effects of wine volatiles on the perception of taste and astringency in wine. Food Chem 121:1139–1149. https://doi.org/10.1016/j.foodchem.2010.01.061
33. Sefton MA, Williams PJ (1991) Generation of oxidation artifacts during the hydrolysis of norisoprenoid glycosides by fungal enzyme preparations. J Agric Food Chem 39:1994–1997
34. Pineau B, Barbe J-C, Van Leeuwen C, Dubourdieu D (2007) Which impact for β-damascenone on red wines aroma? J Agric Food Chem 55:4103–4108
35. Mendes-Pinto MM (2009) Carotenoid breakdown products the—norisoprenoids—in wine aroma. Arch Biochem Biophys 483: 236–245
36. Daniel MA, Elsey GM, Capone DL, Perkins MV, Sefton MA (2004) Fate of damascenone in wine: the role of SO2. J Agric Food Chem 52:8127–8131
37. Kotseridis Y, Baumes RL, Bertrand A, Skouroumounis GK (1999) Quantitative determination of β-ionone in red wines and grapes of Bordeaux using a stable isotope dilution assay. J Chromatogr A 848:317–325
38. Mestres M, Busto O, Guasch J (2000) Analysis of organic sulfur compounds in wine aroma. J Chromatogr A 881:569–581. https://doi.org/10.1016/S0021-9673(00)00220-X
39. Helwi P, Thibon C, Habran A, Hilbert G, Guillaumie S, Delrot S, Darriet P, van Leeuwen C (2015) Effect of vine nitrogen status, grapevine variety and rootstock on the levels of berry S-glutathionylated and S-cysteinylated precursors of 3-sulfanylhexan-1-ol. OENO One 49:253–265. https://doi.org/10.20870/oeno-one.2015.49.4.40
40. Carlin S, Piergiovanni M, Pittari E, Tiziana Lisanti M, Moio L, Piombino P, Marangon M, Curioni A, Rolle L, Río Segade S et al (2022) The contribution of varietal thiols in the diverse aroma of Italian monovarietal white wines. Food Res Int 157:

111404. https://doi.org/10.1016/j.foodres.2022.111404

41. Roland A, Schneider R, Razungles A, Cavelier F (2011) Varietal thiols in wine: discovery, analysis and applications. Chem Rev 111: 7355–7376. https://doi.org/10.1021/cr100205b
42. Waterhouse AL, Sacks GL, Jeffery DW (2016) Understanding wine chemistry. Wiley, New York
43. Howell KS, Swiegers JH, Elsey GM, Siebert TE, Bartowsky EJ, Fleet GH, Pretorius IS, de MA (2004) Variation in 4-mercapto-4-methyl-pentan-2-one release by Saccharomyces cerevisiae commercial wine strains. FEMS Microbiol Lett 240:125–129. https://doi.org/10.1016/j.femsle.2004.09.022
44. Tominaga T, Masneuf-Pomarède I, Dubourdieu D (1995) A S-cysteine conjugate, precursor of aroma of White Sauvignon. OENO One 29:227–232. https://doi.org/10.20870/oeno-one.1995.29.4.1123
45. Rauhut D (2017) Usage and formation of sulphur compounds. In: König H, Unden G, Fröhlich J (eds) Biology of microorganisms on grapes, in must and in wine. Springer International Publishing, Cham, pp 255–291
46. Darriet P, Tominaga T, Lavigne V, Boidron J-N, Dubourdieu D (1995) Identification of a powerful aromatic component of Vitis vinifera L. var. sauvignon wines: 4-mercapto-4-methylpentan-2-one. Flavour Fragrance J 10:385–392. https://doi.org/10.1002/ffj.2730100610
47. Cerreti M, Esti M, Benucci I, Liburdi K, de Simone C, Ferranti P (2015) Evolution of S-cysteinylated and S-glutathionylated thiol precursors during grape ripening of Vitis vinifera L. cvs Grechetto, Malvasia del Lazio and Sauvignon Blanc. Aust J Grape Wine Res 21:411–416. https://doi.org/10.1111/ajgw.12152
48. Choné X, Trégoat O, Leeuwen CV, Dubourdieu D (2000) Vine water deficit: among the 3 applications of pressure chamber, stem water potential is the most sensitive indicator. J Int des Sci de la Vigne et du Vin 34:169–176
49. Helwi P, Guillaumie S, Thibon C, Keime C, Habran A, Hilbert G, Gomes E, Darriet P, Delrot S, van Leeuwen C (2016) Vine nitrogen status and volatile thiols and their precursors from plot to transcriptome level. BMC Plant Biol 16:173. https://doi.org/10.1186/s12870-016-0836-y
50. Román T, Tonidandel T, Larcher R, Celotti E, Nicolini G (2018) Importance of polyfunctional thiols on semi-industrial Gewürztraminer wines and the correlation to technological treatments. Eur Food Res Technol 244:379–386. https://doi.org/10.1007/s00217-017-2963-6
51. Cordente AG, Curtin CD, Varela C, Pretorius IS (2012) Flavour-active wine yeasts. Appl Microbiol Biotechnol 96:601–618. https://doi.org/10.1007/s00253-012-4370-z
52. Frost R, Quiñones I, Veldhuizen M, Alava JI, Small D, Carreiras M (2015) What can the brain teach us about winemaking? An fMRI study of alcohol level preferences. PLoS One 10:e0119220. https://doi.org/10.1371/journal.pone.0119220
53. Ruiz J, Belda I, Beisert B, Navascués E, Marquina D, Calderón F, Rauhut D, Santos A, Benito S (2018) Analytical impact of Metschnikowia pulcherrima in the volatile profile of Verdejo white wines. Appl Microbiol Biotechnol 102:8501–8509. https://doi.org/10.1007/s00253-018-9255-3
54. Dunlevy JD, Kalua CM, Keyzers RA, Boss PK (2009) The production of flavour & aroma compounds in grape berries. In: Roubelakis-Angelakis KA (ed) Grapevine molecular physiology & biotechnology. Springer, Dordrecht, pp 293–340
55. Sidhu D, Lund J, Kotseridis Y, Saucier C (2015) Methoxypyrazine analysis and influence of viticultural and enological procedures on their levels in grapes, musts, and wines. Crit Rev Food Sci Nutr 55:485–502
56. Koch A, Doyle CL, Matthews MA, Williams LE, Ebeler SE (2010) 2-Methoxy-3-isobutyl-pyrazine in grape berries and its dependence on genotype. Phytochemistry 71:2190–2198. https://doi.org/10.1016/j.phytochem.2010.09.006
57. Ward S, Petrie P, Johnson T, Boss P, Bastian S (2015) unripe berries and petioles in Vitis vinifera cv. Cabernet Sauvignon fermentations impact sensory and chemical profiles. Am J Enol Vitic 66:435. https://doi.org/10.5344/ajev.2015.15016
58. Roujou de Boubée D, Van Leeuwen C, Dubourdieu D (2000) Organoleptic impact of 2-methoxy-3-isobutylpyrazine on red bordeaux and loire wines. Effect of environmental conditions on concentrations in grapes during ripening. J Agric Food Chem 48: 4830–4834. https://doi.org/10.1021/jf000181o
59. Zhao X, Ju Y, Wei X, Dong S, Sun X, Fang Y (2019) Significance and transformation of 3-alkyl-2-methoxypyrazines through grapes to wine: Olfactory properties, metabolism, biochemical regulation, and the HP–MP cycle. Molecules 24:4598

60. González Pereira A, Corral M, García Oliveira P, Rodríguez M, Jiménez-López C, Lourenço-Lopes C, Barros L, Ferreira I, Prieto M, Simal-Gandara J (2020) Management of wine aroma compounds: principal basis and future perspectives. In: Winemaking: stabilization, aging chemistry and biochemistry. IntechOpen
61. Ryona I, Leclerc S, Sacks GL (2010) Correlation of 3-isobutyl-2-methoxypyrazine to 3-isobutyl-2-hydroxypyrazine during maturation of bell pepper (Capsicum annuum) and wine grapes (Vitis vinifera). J Agric Food Chem 58:9723–9730. https://doi.org/10.1021/jf102072w
62. Hashizume K, Umeda N (1996) Methoxypyrazine content of Japanese red wines. Biosci Biotechnol Biochem 60:802–805
63. Mendez-Costabel MP, Wilkinson KL, Bastian SEP, Jordans C, McCarthy M, Ford CM, Dokoozlian NK (2014) Effect of increased irrigation and additional nitrogen fertilisation on the concentration of green aroma compounds in Vitis vinifera L. Merlot fruit and wine. Aust J Grape Wine Res 20:80–90. https://doi.org/10.1111/ajgw.12062
64. Rajchl A, Čížková H, Voldřich M, Lukešová D, Panovska Z (2009) Methoxypyrazines in Sauvignon blanc wines, detection of addition of artificial aroma. Czech J Food Sci 27:259–266
65. Seifert RM, Buttery RG, Guadagni DG, Black DR, Harris JG (1970) Synthesis of some 2-methoxy-3-alkylpyrazines with strong bell pepper-like odors. J Agric Food Chem 18: 246–249. https://doi.org/10.1021/jf60168a002
66. Maarse H (1992) Wine, distilled beverages. Volatile compounds in foods and beverages. Arcel Dekker Inc, Amsterdam, pp 483–580
67. Ferreira V, López R, Cacho JF (2000) Quantitative determination of the odorants of young red wines from different grape varieties. J Sci Food Agric 80:1659–1667. https://doi.org/10.1002/1097-0010(20000901)80:11
68. Franco M, Peinado RA, Medina M, Moreno J (2004) Off-vine grape drying effect on volatile compounds and aromatic series in must from Pedro Ximénez grape variety. J Agric Food Chem 52:3905–3910. https://doi.org/10.1021/jf0354949
69. Wood C, Siebert TE, Parker M, Capone DL, Elsey GM, Pollnitz AP, Eggers M, Meier M, Vössing T, Widder S et al (2008) From wine to pepper: rotundone, an obscure sesquiterpene, is a potent spicy aroma compound. J Agric Food Chem 56:3738–3744. https://doi.org/10.1021/jf800183k
70. Darriet P, Thibon C, Dubourdieu D (2012) Aroma and aroma precursors in grape berry. Biochem Grape Berry:111–136. https://doi.org/10.2174/978160805360511201010111
71. Fleet GH (2008) Wine yeasts for the future. FEMS Yeast Res 8:979–995. https://doi.org/10.1111/j.1567-1364.2008.00427.x
72. Garde-Cerdán T, Ancín-Azpilicueta C (2006) Contribution of wild yeasts to the formation of volatile compounds in inoculated wine fermentations. Eur Food Res Technol 222:15–25
73. Ferreira V, Lopez R (2019) The actual and potential aroma of winemaking grapes. Biomol Ther 9. https://doi.org/10.3390/biom9120818
74. Gonzalez R, Morales P (2022) Truth in wine yeast. Microb Biotechnol 15:1339–1356. https://doi.org/10.1111/1751-7915.13848
75. Maicas S (2021) Advances in wine fermentation. Fermentation 7:187
76. Pretorius IS (2019) Tasting the terroir of wine yeast innovation. FEMS Yeast Res 20. https://doi.org/10.1093/femsyr/foz084
77. Jolly NP, Varela C, Pretorius IS (2014) Not your ordinary yeast: non-Saccharomyces yeasts in wine production uncovered. FEMS Yeast Res 14:215–237. https://doi.org/10.1111/1567-1364.12111
78. Tufariello M, Fragasso M, Pico J, Panighel A, Castellarin SD, Flamini R, Grieco F (2021) Influence of non-saccharomyces on wine chemistry: a focus on aroma-related compounds. Molecules 26. https://doi.org/10.3390/molecules26030644
79. Romano P, Braschi G, Siesto G, Patrignani F, Lanciotti R (1921) Role of yeasts on the sensory component of wines. Foods 2022:11
80. Querol A, Fleet GH (2006) Food and beverage spoilage yeasts. In: Querol A, Fleet G (eds) Yeasts in food and beverages. Springer, Berlin
81. Agnolucci M, Tirelli A, Cocolin L, Toffanin A (2017) Brettanomyces bruxellensis yeasts: impact on wine and winemaking. World J Microbiol Biotechnol 33:180. https://doi.org/10.1007/s11274-017-2345-z
82. Carrau F, Gaggero C, Aguilar PS (2015) Yeast diversity and native vigor for flavor phenotypes. Trends Biotechnol 33:148–154. https://doi.org/10.1016/j.tibtech.2014.12.009

83. Ciani M, Comitini F, Mannazzu I, Domizio P (2010) Controlled mixed culture fermentation: a new perspective on the use of non-Saccharomyces yeasts in winemaking. FEMS Yeast Res 10:123–133. https://doi.org/10.1111/j.1567-1364.2009.00579.x
84. Bisson LF (2019) Gene expression in yeasts during wine fermentation. In: Romano P, Ciani M, Fleet GH (eds) Yeasts in the production of wine. Springer, New York, pp 165–209
85. Anfang N, Brajkovich M, Goddard MR (2009) Co-fermentation with Pichia kluyveri increases varietal thiol concentrations in Sauvignon Blanc. Aust J Grape Wine Res 15:1–8
86. Belda I, Navascués E, Marquina D, Santos A, Calderon F, Benito S (2015) Dynamic analysis of physiological properties of Torulaspora delbrueckii in wine fermentations and its incidence on wine quality. Appl Microbiol Biotechnol 99:1911–1922. https://doi.org/10.1007/s00253-014-6197-2
87. Bely M, Stoeckle P, Masneuf-Pomarède I, Dubourdieu D (2008) Impact of mixed Torulaspora delbrueckii-Saccharomyces cerevisiae culture on high-sugar fermentation. Int J Food Microbiol 122:312–320. https://doi.org/10.1016/j.ijfoodmicro.2007.12.023
88. Benito S (2018) The impact of Torulaspora delbrueckii yeast in winemaking. Appl Microbiol Biotechnol 102:3081–3094. https://doi.org/10.1007/s00253-018-8849-0
89. Benito S (2018) The impacts of Lachancea thermotolerans yeast strains on winemaking. Appl Microbiol Biotechnol 102:6775–6790. https://doi.org/10.1007/s00253-018-9117-z
90. Benito S (2020) Combined use of lachancea thermotolerans and schizosaccharomyces pombe in winemaking: a review. Microorganisms 8. https://doi.org/10.3390/microorganisms8050655
91. Blanco P, Rabuñal E, Neira N, Castrillo D (2020) Dynamic of lachancea thermotolerans population in monoculture and mixed fermentations: impact on wine characteristics. Beverages 6:36
92. Ciani M (1995) Continuous deacidification of wine by immobilized Schizosaccharomyces pombe cells: evaluation of malic acid degradation rate and analytical profiles. J Appl Microbiol 79:631–634
93. Clemente-Jimenez J, Mingorance-Cazorla L, Martínez-Rodríguez S, Las Heras-Vázquez F, Rodríguez-Vico F (2005) Influence of sequential yeast mixtures on wine fermentation. Int J Food Microbiol 98:301–308
94. Comitini F, Gobbi M, Domizio P, Romani C, Lencioni L, Mannazzu I, Ciani M (2011) Selected non-Saccharomyces wine yeasts in controlled multistarter fermentations with Saccharomyces cerevisiae. Food Microbiol 28:873–882
95. Ferraro L, Fatichenti F, Ciani M (2000) Pilot scale vinification process using immobilized Candida stellata cells and Saccharomyces cerevisiae. Process Biochem 35:1125–1129
96. Garcia A, Carcel C, Dulau L, Samson A, Aguera E, Agosin E, Günata Z (2002) Influence of a mixed culture with Debaryomyces vanriji and Saccharomyces cerevisiae on the volatiles of a Muscat wine. J Food Sci 67: 1138–1143
97. Giorello F, Valera MJ, Martin V, Parada A, Salzman V, Camesasca L, Fariña L, Boido E, Medina K, Dellacassa E et al (2019) Genomic and transcriptomic basis of hanseniaspora vineae's impact on flavor diversity and wine quality. Appl Environ Microbiol 85. https://doi.org/10.1128/aem.01959-18
98. Ivit NN, Longo R, Kemp B (2020) The effect of non-Saccharomyces and Saccharomyces non-cerevisiae yeasts on ethanol and glycerol levels in wine. Fermentation 6:77
99. Jolly N, Augustyn O, Pretorius I (2003) The use of Candida pulcherrima in combination with Saccharomyces cerevisiae for the production of Chenin blanc wine. South African J Enol Vitic 24:63–69
100. Kapsopoulou K, Mourtzini A, Anthoulas M, Nerantzis E (2007) Biological acidification during grape must fermentation using mixed cultures of Kluyveromyces thermotolerans and Saccharomyces cerevisiae. World J Microbiol Biotechnol 23:735–739
101. Kim D-H, Hong Y-A, Park H-D (2008) Co-fermentation of grape must by Issatchenkia orientalis and Saccharomyces cerevisiae reduces the malic acid content in wine. Biotechnol Lett 30:1633–1638
102. Medina K, Boido E, Fariña L, Gioia O, Gomez ME, Barquet M, Gaggero C, Dellacassa E, Carrau F (2013) Increased flavour diversity of Chardonnay wines by spontaneous fermentation and co-fermentation with Hanseniaspora vineae. Food Chem 141:2513–2521. https://doi.org/10.1016/j.foodchem.2013.04.056
103. Mendoza LM, de Nadra MCM, Farías ME (2007) Kinetics and metabolic behavior of a composite culture of Kloeckera apiculata and Saccharomyces cerevisiae wine related strains. Biotechnol Lett 29:1057–1063

104. Morata A, Escott C, Bañuelos MA, Loira I, Del Fresno JM, González C, Suárez-Lepe JA (2019) Contribution of non-Saccharomyces yeasts to wine freshness: a review. Biomolecules 10:34
105. Moreira N, Mendes F, de Pinho PG, Hogg T, Vasconcelos I (2008) Heavy sulphur compounds, higher alcohols and esters production profile of Hanseniaspora uvarum and Hanseniaspora guilliermondii grown as pure and mixed cultures in grape must. Int J Food Microbiol 124:231–238
106. Moreira N, Pina C, Mendes F, Couto JA, Hogg T, Vasconcelos I (2011) Volatile compounds contribution of Hanseniaspora guilliermondii and Hanseniaspora uvarum during red wine vinifications. Food Control 22:662–667. https://doi.org/10.1016/j.foodcont.2010.07.025
107. Palomero F, Morata A, Benito S, Calderón F, Suárez-Lepe J (2009) New genera of yeasts for over-lees aging of red wine. Food Chem 112:432–441
108. Porter TJ, Divol B, Setati ME (2019) Lachancea yeast species: origin, biochemical characteristics and oenological significance. Food Res Int 119:378–389. https://doi.org/10.1016/j.foodres.2019.02.003
109. Soden A, Francis I, Oakey H, Henschke P (2000) Effects of co-fermentation with Candida stellata and Saccharomyces cerevisiae on the aroma and composition of Chardonnay wine. Aust J Grape Wine Res 6:21–30
110. Toro M, Vazquez F (2002) Fermentation behaviour of controlled mixed and sequential cultures of Candida cantarellii and Saccharomyces cerevisiae wine yeasts. World J Microbiol Biotechnol 18:351–358
111. Varela C, Siebert T, Cozzolino D, Rose L, McLean H, Henschke PA (2009) Discovering a chemical basis for differentiating wines made by fermentation with 'wild' indigenous and inoculated yeasts: role of yeast volatile compounds. Aust J Grape Wine Res 15:238–248. https://doi.org/10.1111/j.1755-0238.2009.00054.x
112. Vicente J, Calderón F, Santos A, Marquina D, Benito S (2021) High potential of pichia kluyveri and other pichia species in wine technology. Int J Mol Sci 22:1196
113. Vicente J, Ruiz J, Belda I, Benito-Vázquez I, Marquina D, Calderón F, Santos A, Benito S (2020) The genus metschnikowia in enology. Microorganisms 8:1038. https://doi.org/10.3390/microorganisms8071038
114. Vilela A (2018) Lachancea thermotolerans, the non-Saccharomyces yeast that reduces the volatile acidity of wines. Fermentation 4:56
115. Kalua CM, Boss PK (2010) Comparison of major volatile compounds from Riesling and Cabernet Sauvignon grapes (Vitis vinifera L.) from fruitset to harvest. Aust J Grape Wine Res 16:337–348. https://doi.org/10.1111/j.1755-0238.2010.00096.x
116. Oliveira JM, Faria M, Sá F, Barros F, Araújo IM (2006) C 6-alcohols as varietal markers for assessment of wine origin. Anal Chim Acta 563:300–309
117. Reader HP, Dominguez M (2003) Fortified wines sherry, port and madeira. In: Fermented beverage production, p 157–194, doi: https://doi.org/10.1007/978-1-4615-0187-9_8
118. Oliveira e Silva H, de Pinho PG, Machado BP, Hogg T, Marques JC, Camara JS, Albuquerque F, Ferreira AC (2008) Impact of forced-aging process on madeira wine flavor. J Agric Food Chem 56:11989–11996
119. Pereira V, Albuquerque FM, Ferreira AC, Cacho J, Marques JC (2011) Evolution of 5-hydroxymethylfurfural (HMF) and furfural (F) in fortified wines submitted to overheating conditions. Food Res Int 44:71–76. https://doi.org/10.1016/j.foodres.2010.11.011
120. Câmara JS, Alves MA, Marques JCC, Liu M, Zeng Z, Tian Y, Fitzgerald G, James KJ, MacNamara K, Stack MA et al (2020) Changes in volatile composition of Madeira wines during their oxidative ageing. J Chromatogr A 563:188–197. https://doi.org/10.1016/j.cclet.2007.06.013
121. Perestrelo R, Barros ASS, Câmara JSS, Rocha SMM (2011) In-depth search focused on furans, lactones, volatile phenols, and acetals as potential age markers of Madeira wines by comprehensive two-dimensional gas chromatography with time-of-flight mass spectrometry combined with solid phase microextraction. J Agric Food Chem 59:3186–3204
122. Abreu T, Perestrelo R, Bordiga M, Locatelli M, Daniel Coisson J, Camara JS (2021) The flavor chemistry of fortified wines-a comprehensive approach. Foods 10. https://doi.org/10.3390/foods10061239
123. Călugăr A, Coldea TE, Pop CR, Pop TI, Babeş AC, Bunea CI, Manolache M, Gal E (2020) Evaluation of volatile compounds during ageing with oak chips and oak barrel of Muscat Ottonel wine. PRO 8:1000
124. Campo E, Ferreira V, Escudero A, Marqués JC, Cacho J (2006) Quantitative gas

chromatography-olfactometry and chemical quantitative study of the aroma of four Madeira wines. Anal Chim Acta 563:180–187
125. Perestrelo R, Silva C, Câmara JS, Maria R, Perestrelo S, Câmara JS (2019) Madeira wine volatile profile: a platform to establish madeira wine aroma descriptors. Molecules 24:3028
126. Perestrelo R, Jaouhari Y, Abreu T, Castillo MM, Travaglia F, Pereira JAM, Câmara JS, Bordiga M (2023) The fingerprint of fortified wines–from the sui generis production processes to the distinctive aroma. Foods 12: 2558
127. Perestrelo R, Silva CL, Silva P, Câmara JS (2018) Impact of storage time and temperature on volatomic signature of Tinta Negra wines by LLME/GC-ITMS. Food Res Int 109:99–111
128. Pereira AC, Reis MS, Saraiva PM, Marques JC (2010) Aroma ageing trends in GC/MS profiles of liqueur wines. Anal Chim Acta 659: 93–101. https://doi.org/10.1016/j.aca.2009.11.060
129. Pereira V, Cacho J, Marques JC (2014) Volatile profile of Madeira wines submitted to traditional accelerated ageing. Food Chem 162:122–134. https://doi.org/10.1016/j.foodchem.2014.04.039
130. Gaspar JM, Freitas AI, Zhao Q, Leça JM, Pereira V, Marques JC (2019) Is sotolon relevant to the aroma of madeira wine blends? Biomol Ther 9:E720–E729
131. da Silva Ferreira AC, Barbe JC, Bertrand A (2002) Heterocyclic acetals from glycerol and acetaldehyde in Port wines: evolution with aging. J Agric Food Chem 50:2560–2564
132. Perestrelo R, Silva C, Gonçalves C, Castillo M, Câmara JS (2020) An approach of the madeira wine chemistry. Beverages 6: 12. https://doi.org/10.3390/beverages6010012
133. Alañón ME, Rubio H, Díaz-Maroto MC, Pérez-Coello MS (2010) Monosaccharide anhydrides, new markers of toasted oak wood used for ageing wines and distillates. Food Chem 119:505–512
134. Ortu E, Caboni P (2017) Levels of 5-hydroxymethylfurfural, furfural, 2-furoic acid in sapa syrup, Marsala wine and bakery products. Int J Food Prop 20:S2543–S2551
135. Câmara JS, Marques JC, Alves MA, Silva Ferreira AC (2004) 3-Hydroxy-4,5-dimethyl-2 (5H)-furanone levels in fortified Madeira wines: relationship to sugar content. J Agric Food Chem 52:6765–6769
136. Kheir J, Salameh D, Strehaiano P, Brandam C, Lteif R (2013) Impact of volatile phenols and their precursors on wine quality and control measures of Brettanomyces/Dekkera yeasts. Eur Food Res Technol 237:655–671
137. Ayestaran B, Martinez-Lapuente L, Guadalupe Z, Canals C, Adell E, Vilanova M (2019) Effect of the winemaking process on the volatile composition and aromatic profile of Tempranillo Blanco wines. Food Chem 276:187–194. https://doi.org/10.1016/j.foodchem.2018.10.013
138. Vilanova M, Fernandez-Fernandez E, Yuste J (2020) Abiotic stress management and Verdejo wine composition: cluster thinning in different water regimes. J Sci Food Agric 100:1515–1523. https://doi.org/10.1002/jsfa.10159
139. Fairbairn S, McKinnon A, Musarurwa HT, Ferreira AC, Bauer FF (2017) The impact of single amino acids on growth and volatile aroma production by saccharomyces cerevisiae strains. Front Microbiol 8:2554. https://doi.org/10.3389/fmicb.2017.02554
140. Rossi L, Foschi M, Biancolillo A, Maggi MA, D'Archivio AA (2023) Optimization of HS-SPME-GC/MS analysis of wine volatiles supported by chemometrics for the aroma profiling of Trebbiano d'Abruzzo and pecorino white wines produced in Abruzzo (Italy). Molecules 28:1534
141. Perestrelo R, Silva CL, Silva P, Câmara JS (2018) Establishment of the volatile signature of wine-based aromatic vinegars subjected to maceration. Molecules 23. https://doi.org/10.3390/molecules23020499
142. Torrens J, Riu-Aumatell M, López-Tamames E, Buxaderas S (2004) Volatile compounds of red and white wines by headspace-solid-phase microextraction using different fibers. J Chromatogr Sci 42:310–316. https://doi.org/10.1093/chromsci/42.6.310
143. Nguyen ANH, Capone DL, Johnson TE, Jeffery DW, Danner L, Bastian SEP (2019) Volatile composition and sensory profiles of a shiraz wine product made with pre- and post-fermentation additions of ganoderma lucidum extract. Foods 8. https://doi.org/10.3390/foods8110538
144. Román SM-S, Garde-Cerdán T, Baroja E, Rubio-Bretón P, Pérez-Álvarez EP (2020) Foliar application of phenylalanine plus methyl jasmonate as a tool to improve Grenache grape aromatic composition. Sci Hortic 272:109515. https://doi.org/10.1016/j.scienta.2020.109515

145. de Castilhos MBM, Del Bianchi VL, Gomez-Alonso S, Garcia-Romero E, Hermosin-Gutierrez I (2019) Sensory descriptive and comprehensive GC-MS as suitable tools to characterize the effects of alternative wine-making procedures on wine aroma. Part I: BRS Carmem and BRS Violeta. Food Chem 272:462–470. https://doi.org/10.1016/j.foodchem.2018.08.066

146. Hasan CK, Ghiasvand A, Lewis TW, Nesterenko PN, Paull B (2020) Recent advances in stir-bar sorptive extraction: coatings, technical improvements, and applications. Anal Chim Acta 1139:222–240. https://doi.org/10.1016/j.aca.2020.08.021

147. He M, Wang Y, Zhang Q, Zang L, Chen B, Hu B (2021) Stir bar sorptive extraction and its application. J Chromatogr A 1637:461810. https://doi.org/10.1016/j.chroma.2020.461810

148. Marín-San Román S, Carot-Sierra JM, Sáenz de Urturi I, Rubio-Bretón P, Pérez-Álvarez EP, Garde-Cerdán T (2022) Optimization of stir bar sorptive extraction (SBSE) and multi-stir bar sorptive extraction (mSBSE) to improve must volatile compounds extraction. LWT 172:114182. https://doi.org/10.1016/j.lwt.2022.114182

149. Rios-Reina R, Segura-Borrego MP, Garcia-Gonzalez DL, Morales ML, Callejon RM (2019) A comparative study of the volatile profile of wine vinegars with protected designation of origin by headspace stir bar sorptive extraction. Food Res Int 123:298–310. https://doi.org/10.1016/j.foodres.2019.04.071

150. Hauser B, Popp P (2001) Membrane-assisted solvent extraction of organochlorine compounds in combination with large-volume injection/gas chromatography-electron capture detection. J Sep Sci 24:551–560. https://doi.org/10.1002/1615-9314(20010801)

151. Pisaniello L, Watson F, Siebert T, Francis L, Hixson JL (2022) The varietal influence of flavour precursors from grape marc on monoterpene and C13-norisoprenoid profiles in wine as determined by membrane-assisted solvent extraction (MASE) GC-MS. Molecules 27. https://doi.org/10.3390/molecules27072046

152. Câmara JS, Lourenço S, Silva C, Lopes A, Andrade C, Perestrelo R (2020) Exploring the potential of wine industry by-products as source of additives to improve the quality of aquafeed. Microchem J 155:104758. https://doi.org/10.1016/j.microc.2020.104758

153. Slaghenaufi D, Guardini S, Tedeschi R, Ugliano M (2019) Volatile terpenoids, norisoprenoids and benzenoids as markers of fine scale vineyard segmentation for Corvina grapes and wines. Food Res Int 125:108507. https://doi.org/10.1016/j.foodres.2019.108507

154. Picariello L, Slaghenaufi D, Ugliano M (2020) Fermentative and post-fermentative oxygenation of Corvina red wine: influence on phenolic and volatile composition, colour and wine oxidative response. J Sci Food Agric 100:2522–2533. https://doi.org/10.1002/jsfa.10278

155. Chen L, Darriet P (2022) Qualitative screening of volatile thiols in wine by selective silver ion solid-phase extraction with heart-cutting multidimensional gas chromatography mass spectrometry/olfactometry. J Agric Food Chem 70:4701–4711. https://doi.org/10.1021/acs.jafc.2c00243

156. Jasmins G, Perestrelo R, Coïsson JD, Sousa P, Teixeira JA, Bordiga M, Câmara JS (2023) Tracing the volatilomic fingerprint of the most popular italian fortified wines. Foods 12. https://doi.org/10.3390/foods12102058

157. Guerrero-Chanivet M, Valcárcel-Muñoz MJ, García-Moreno MV, Guillén-Sánchez DA (2020) Characterization of the aromatic and phenolic profile of five different wood chips used for ageing spirits and wines. Foods 9:1613

158. Ivanova-Petropulos V, Petruševa D, Mitrev S (2020) Rapid and simple method for determination of target organic acids in wine using HPLC-DAD analysis. Food Anal Methods 13:1078–1087. https://doi.org/10.1007/s12161-020-01724-4

159. Coelho EM, da Silva Padilha CV, Miskinis GA, de Sá AGB, Pereira GE, de Azevêdo LC, dos Santos Lima M (2018) Simultaneous analysis of sugars and organic acids in wine and grape juices by HPLC: method validation and characterization of products from northeast Brazil. J Food Compos Anal 66:160–167. https://doi.org/10.1016/j.jfca.2017.12.017

160. Vievard J, Amoikon TL-S, Coulibaly NA, Devouge-Boyer C, Arellano-Sánchez MG, Aké MFD, Djeni NDT, Mignot M (2022) Extraction and quantification of pesticides and metals in palm wines by HS-SPME/GC–MS and ICP-AES/MS. Food Chem 393:133352. https://doi.org/10.1016/j.foodchem.2022.133352

161. Lisanti MT, Laboyrie J, Marchand-Marion S, de Revel G, Moio L, Riquier L, Franc C

(2021) Minty aroma compounds in red wine: development of a novel automated HS-SPME-arrow and gas chromatography-tandem mass spectrometry quantification method. Food Chem 361:130029. https://doi.org/10.1016/j.foodchem.2021.130029

162. Ju YL, Xu XL, Yu YK, Liu M, Wang WN, Wu JR, Liu BC, Zhang Y, Fang YL (2023) Effects of winemaking techniques on the phenolics, organic acids, and volatile compounds of Muscat wines. Food Biosci 54:102937. https://doi.org/10.1016/j.fbio.2023.102937

163. Lai Y-T, Chen C-H, Lo Y-C, Hsieh C-W, Hsu F-C, Cheng K-C (2023) Application of aroma-producing yeasts and ageing technology in Kyoho-fortified wine. Eur Food Res Technol 249:2849. https://doi.org/10.1007/s00217-023-04332-8

164. Piergiovanni M, Carlin S, Lotti C, Vrhovsek U, Mattivi F (2023) Development of a fully automated method HS-SPME-GC-MS/MS for the determination of odor-active carbonyls in wines: a "Green" approach to improve robustness and productivity in the oenological analytical chemistry. J Agric Food Chem. https://doi.org/10.1021/acs.jafc.2c07083

165. Vakinti M, Mela S-M, Fernández E, Psillakis E (2019) Room temperature and sensitive determination of haloanisoles in wine using vacuum-assisted headspace solid-phase microextraction. J Chromatogr A 1602:142–149. https://doi.org/10.1016/j.chroma.2019.03.047

166. Šorgić S, Sredović Ignjatović I, Antić M, Šaćirović S, Pezo L, Čejić V, Đurović S (2022) Monitoring of the wines' quality by gas chromatography: hss-gc/fid method development, validation, verification, for analysis of volatile compounds. Fermentation 8:38

167. Lenti L, Nartea A, Orhotohwo OL, Pacetti D, Fiorini D (2022) Development and validation of a new GC-FID method for the determination of short and medium chain free fatty acids in wine. Molecules 27:27. https://doi.org/10.3390/molecules27238195

168. Qin B, Wang X, Tang L, Wang S, Shi Y, Zhao L, Jiang H (2022) Comparative study of headspace and headspace single drop microextraction combined with GC for the determination of methanol in wine. J Chromatogr A 1673:463079. https://doi.org/10.1016/j.chroma.2022.463079

169. Welke JE, Nicolli KP, Hernandes KC, Biasoto ACT, Zini CA (2022) Adaptation of an olfactometric system in a GC-FID in combination with GCxGC/MS to evaluate odor-active compounds of wine. Food Chem 370:131004. https://doi.org/10.1016/j.foodchem.2021.131004

170. Bordiga M, Rinaldi M, Locatelli M, Piana G, Travaglia F, Coïsson JD, Arlorio M (2013) Characterization of Muscat wines aroma evolution using comprehensive gas chromatography followed by a post-analytic approach to 2D contour plots comparison. Food Chem 140:57–67. https://doi.org/10.1016/j.foodchem.2013.02.051

Chapter 10

In-Mouth Wine Aroma Analysis

María Pérez-Jiménez, Carolina Muñoz González, and María Ángeles Pozo-Bayón

Abstract

In-mouth aroma analysis under wine-tasting conditions is a challenging task. By combining the use of analytical tools such as solid phase micro-extraction (SPME) or stir bar sorptive extraction (SBSE) with the design of different devices and/or methodological approaches, it is possible to monitor the aroma release in the mouth at different times while simulating the consumption of wine. Two recently employed techniques to monitor in-mouth aroma release from wine are intra-oral SPME and in-mouth headspace sorptive extraction (HSSE). These procedures consist in the intra-oral extraction of wine volatiles into adsorbent/absorbent polymers after the oral exposure to the wine by using an SPME fiber or a stir bar also called a twister. Then, aroma compounds contained in both types of polymers are desorbed and analyzed by gas chromatography-mass spectrometry (GC-MS). Using these tools, the total amount of aroma released by each odorant can be determined. Through this, it is possible to compare the intra-oral amounts and release patterns of different types of wine volatiles, which can be useful for determining their oral aroma persistence. This chapter provides a step-by-step guideline for the extraction, desorption, and analysis by GC-MS of the wine aroma compounds released in the mouth after wine tasting by using intra-oral SPME or in-mouth HSSE procedures.

Key words Intra-oral aroma analysis, Aroma release, Retronasal aroma, Aroma persistence, GC-MS, Wine oral processing

1 Introduction

The intra-oral solid phase microextraction (SPME) and the in-mouth headspace sorptive extraction (HSSE) techniques coupled to gas chromatography-mass spectrometry (GC-MS) allowed us to determine the presence of aroma molecules in the mouth after wine tasting [1, 2]. Both techniques are based on the monitoring of the in-mouth headspace immediately after oral exposure to wine by using adsorbent/absorbent polymers, such as SPME fibers or magnetic stir bars also called twisters, normally used for SBSE. Both techniques rely upon considering that during the in-mouth extractions, the oral cavity works as a closed system as the velum tongue is

María Ángeles Pozo-Bayón and Carolina Muñoz González (eds.), *Wine Analysis and Testing Techniques*, Methods and Protocols in Food Science, https://doi.org/10.1007/978-1-0716-3650-3_10,

closed, and thus, there is no airflow circulation between the nasal and the oral cavities [3]. Additionally, these techniques allow us to monitor the headspace of the mouth at different and specific times after wine tasting [4, 5]. Like this, it can be possible to obtain the release kinetics of different aroma molecules in the mouth and determine both the immediate and the prolonged retronasal wine aroma, also called aroma persistence. Both techniques have recently been used in different works to investigate the impact of wine composition [6, 7] and/or individual factors [8, 9] on the release and persistence of different types of wine volatiles.

The use of SPME fibers or twisters allows us in a single step the extraction and concentration of the aroma compounds contained in the breath after wine tasting in the polymers before the GC-MS analysis of odorants. Unlike other in vivo and online methods, such as proton-transfer-reaction mass spectrometry (PTR-MS) or atmospheric-pressure chemical ionization (APCI-MS), in which volatiles released in the breath are detected in the MS at real time, both, intra-oral SPME and in-mouth HSSE methods are *offline* (cumulative) methods based on a previous chromatographic separation of the breath aroma compounds prior to detection. The main advantage of the intra-oral SPME procedure over the HSSE procedure is that SPME is more affordable and easier to apply. Nonetheless, its main weakness is the limited sensitivity of the method, due to the reduced amount of polymer in the fiber. This is why this technique could be mainly used for major wine volatiles of for working in spiked model wines, with a reinforced aroma profile. Another drawback is that the technique is not fully automated, since it requires the manual desorption of the fibers (one-by-one) in the GC inlet, which limit the number of analyses that can be done per day. On the other hand, the main advantage of the in-mouth HSSE procedure is that is more sensitive, due the large amount of polymer in the twisters, which allow us to work with real wines at natural aroma concentration without the necessity to reinforce the wine aroma profile [2]. Furthermore, in case of having a TDU (thermal desorption unit) combined with an autosampler, another advantage of the in-mouth HSSE procedure is that once all the in-mouth aroma extractions have been completed, the twisters with the breath extracts can be automatically desorbed in the TDU, which enable to analyze several samples per day and to work with larger groups of volunteers [5].

As previously described in the literature, the main steps of both the intra-oral SPME and the in-mouth HSSE procedures consist in: (1) monitor the in-mouth headspace before the oral exposure to wine; (2) in-mouth aroma extraction after the oral exposure to wine, (3) desorption from the polymer, and (4) analysis by GC-MS [1, 2].

2 Materials

The laboratory reagents should be of food grade and material exclusively for using in studies with human subjects. Once the in-mouth extractions have been performed, in the subsequent stages of sample processing, is not necessary to use food grade reagents and materials.

2.1 General Laboratory Material and Reagents

Laboratory material: graduated flasks with cups, precision balance, spatula, glass pipettes, test tubes, wine glasses, glass or plastic containers to store stocks.

Reagents: ethanol, aroma standards.

2.2 Volunteers

For the selection of volunteers (*see* **Note 1**), some aspects related to their physiology should be considered, such as not having known illnesses, allergies to wine components or being pregnant, and to be non-smokers.

Before to starting the assays, the experimental procedure must be explained to the volunteers in detail. Like this, volunteers should be informed about the aims and procedures of the study, which will be conducted according to the guidelines of the Declaration of Helsinki. They should provide their written consent before their participation. Additionally, is important to train the volunteers in the correspondent procedure (intra-oral SPME or in-mouth HSSE) before the beginning of the test to obtain the most accurate possible results.

2.3 Wine and Palate Cleansers

For the wine, serve 15 mL of wine or synthetic wine (depending on the study) (*see* **Note 2**) in a wine glass and cover it to prevent the alteration of the wine (*see* **Note 3**).

For the application of the intra-oral SPME procedure, it is recommended to use spiked (aromatized) wines in order to enrich their volatile profile. While, the in-mouth HSSE method can be applied directly to wine for testing its aroma release, thus an aromatization steps is not needed.

In case of working with aromatized wines or synthetic wines, it is recommended that the aromatization step can be done immediately before the experiment. The final concentration of each target aroma compound may vary depending on the aim of the study and on the characteristics of the odorants (*see* **Note 4**). Previous studies have used concentrations between 1 and 4 mg/L in 15 mL of wine (Table 1). For the aromatization, it is recommended that each aroma compound can be individually added to the wine. For that, it is advisable to prepare two stock solutions of each aroma compound separately. The first solution (Solution 1) should have a high concentration (e.g., 1000–1500 mg/L) (*see* **Note 5**). A highly concentrated solution is easy to prepare, as it is quite difficult to

Table 1
Aroma compounds, concentrations, and monitoring times employed in the literature in different studies using the intra-oral SPME method to monitor oral aroma release during wine tasting

Aroma compounds	Concentration	N° of in-mouth samplings	References
Ethyl hexanoate, β-ionone, linalool, guaiacol, β-phenylethanol and isoamyl acetate	0.5, 1, 1.5, or 2 mg/L	1	Esteban-Fernández et al. [1]
Ethyl hexanoate, β-ionone, linalool, guaiacol, β-phenylethanol and isoamyl acetate	1 mg/L	1	Esteban-Fernández et al. [13]
Ethyl hexanoate, β-ionone, linalool, guaiacol, β-phenylethanol and isoamyl acetate	2 mg/L	2 (t1 = 0 min; t2 = 4 min)	Perez-Jiménez et al. [4]
Ethyl butyrate, isoamyl acetate, ethyl pentanoate, ethyl hexanoate, ethyl octanoate, ethyl decanoate	2 mg/L	2 (t1 = 0 min; t2 = 4 min)	Pérez-Jiménez et al. [14]
Ethyl butyrate, isoamyl acetate, ethyl pentanoate, ethyl hexanoate, ethyl octanoate, ethyl decanoate	4 mg/L	2 (t1 = 0 min; t2 = 4 min)	Muñoz-González et al. [15]
Ethyl butyrate, isoamyl acetate, ethyl pentanoate, ethyl hexanoate, ethyl octanoate, ethyl decanoate	4 mg/L	2 (t1 = 0 min; t2 = 4 min)	Muñoz-González et al. [16]

t1 and t2 are the in-mouth sampling times (in min) after wine spat out

weigh very small quantities of aroma compounds. From Solution 1, a second stock solution (Solution 2) with a lower concentration (e.g., between 200 and 300 mg/L) can be prepared. This Solution 2 will be the working solution to spike the wines.

Palate cleansers: prepare in three different bottles a solution of a teaspoon of bicarbonate in mineral water, a pectin solution of 1 g/L in mineral water, and finally mineral water to clean the mouth of the volunteers before each assay and between wine samples (*see* **Note 6**).

2.4 Intra-Oral Wine Aroma Extraction

In case of the intra-oral SPME procedure, DVB/CAR/PDMS (divinylbenzene/Carboxen/polydimethyl siloxane 50/30 μm film thickness, 2 cm length) coated SPME fibers are usually employed for the extraction of aroma compounds in the mouth. If the analysis requires monitoring aroma release at different times after expectoration of the wine, different fibers will be used for each sampling time (one fiber to monitor immediately after rinsing, another fiber to monitor 5 min later, etc.). The number of fibers needed will depend on the number of in-mouth aroma extractions, considering that fibers are manually injected in the GC (*see* **Note 7**).

To monitor the headspace of the mouth, the fibers should be placed in manual holders for SPME. In order to assure that the fiber does not touch the oral surfaces during the intra-oral

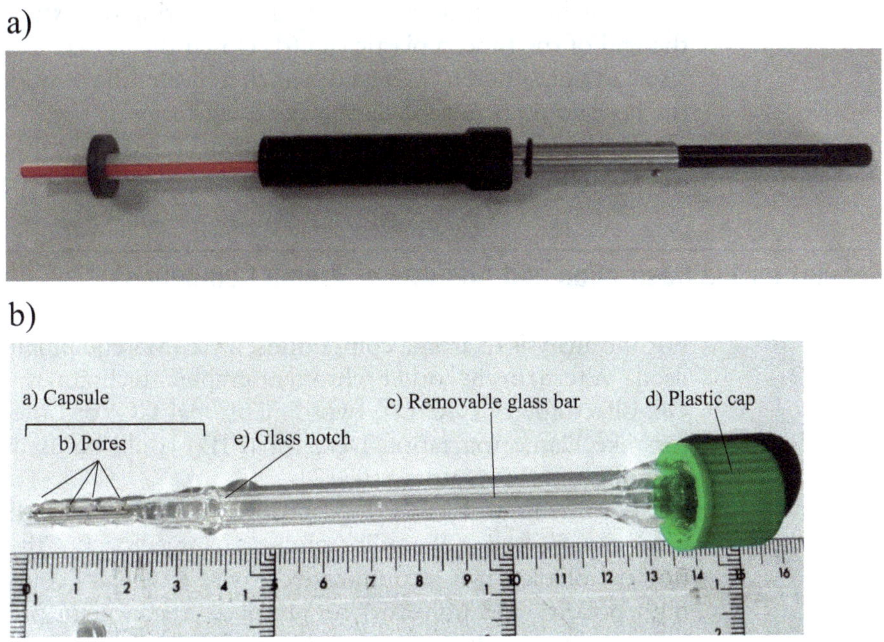

Fig. 1 Devices employed for the intra-oral wine aroma extraction. (**a**) The device employed for the intra-oral SPME. (**b**) The glass device employed for the in-mouth HSSE procedure. (From Perez-Jimenez and Pozo-Bayon [2])

monitorization, a home-made plastic adaptor can be placed above the fiber [1]. This can be a plastic/Teflon tube (like a piece of a drinking straw) with a mark in one of its sides to place the lips always in the same position (Fig. 1a). Plastic tubes should be discarded after the test. Each volunteer might use his/her own adaptor throughout the study.

In the case of the in-mouth HSSE procedure, PDMS twisters of 20 mm length × 0.5 mm thickness are usually employed for the extraction of aroma compounds in the mouth. Different twisters should be employed for each in-mouth extraction. The number of twisters needed for the experiment will depend on the number of in-mouth aroma extractions and on the number of analyses to be performed per day, considering that twisters can be automatically desorbed in the GC (*see* **Note 7**).

For the in-mouth headspace twister extraction, it is necessary to place the twister in a holder device [2]. For instance, in Perez-Jimenez and Pozo-Bayón (2019) they used a tailored made glass holder device developed by Segainvex-UAM (Madrid, Spain) (Fig. 1b). This was a hollow glass tube (14.5 cm long) with a capsule (a) at one end into which the twister was placed. The capsule was homogeneously perforated with eight pores (2–3 mm diameter) (b). The dimensions of the capsule (27 mm long × 5 mm diameter) prevent the movement of the twister during the in-mouth aroma extraction. A removable glass bar (c) was placed inside the hollow tube to prevent air flow during the extraction. At the end of the tube, a plastic cap (d) closed the device. The device also had a glass notch (e) outside which indicates the position where the lips should be placed during the extraction.

For handling the twisters without touching them, a magnet bar can be used.

3 Equipment for the Desorption and Analysis of Aroma Compounds

For the analysis of aroma compounds, a GC-MS equipment can be used. Alternatively, other chromatographic techniques such as GC-olfactometry (GC-O), two-dimensional GC, or other detectors like Flame Ionization Detector (FID) could be used for the identification of volatiles [10–12].

For the intra-oral SPME procedure, the SPME fiber should be manually injected in the split/splitless GC injector. For the separation of volatiles, it is recommended to use a capillary column with high polarity and polyethylene glycol as a stationary phase. For instance, a DB-WAX column (Agilent, j&WScientific, Folsom, CA, USA) with dimensions of 60 m × 0.25 mm and film thickness of 0.50 μm [1]. Nonetheless depending on the type of volatiles of interest other columns might be also used. The carrier gas is usually Helium.

For the in-mouth HSSE procedure, the automated injection of twisters should be done in the autosampler of the GC that must contain the injector adaptor for twisters. Additionally, a thermal desorption unit (TDU) in combination with a CIS-4 (cooled injection system) injector is required. This system allows the thermal desorption of volatiles in the TDU first and then, cryo-focusing the analytes in the CIS-4 system using low temperature (e.g., liquid Nitrogen) prior to their transfer onto the analytical column. For the separation of volatiles, the same column and chromatographic conditions recommended for intra-oral SPME can be used.

4 Methods

It is recommended to perform each analysis at least three times with each volunteer (*see* **Note 8**). The volunteers must be instructed not to drink, eat or smoke 2 h before the assay.

Fifteen minutes before each experiment, the volunteers must perform vigorous rinses with the palate cleanser solutions in the order: bicarbonate solution, pectin solution, and mineral water, to have the most similar oral conditions among them when starting the assay.

The volunteers must introduce the 15 mL of the wine or aromatized wine in the mouth in a single zip doing soft rinses during 30 s (*see* **Note 9**). During rinsing, the lips must be closed and swallowing is not allowed, in order to avoid opening the velum–tongue border prior to expectoration. After rinsing, volunteers should spit out the wine, perform a single swallowing of the remaining saliva in the mouth and wait for 5 s until the first intra-oral aroma extraction.

5 Intra-Oral Wine Aroma Extraction

5.1 Intra-Oral SPME

A schematic representation of the intra-oral SPME procedure is shown in (Fig. 2a).

Once the wine sample has been expectorated and the remaining saliva swallowed, place the SMPE fiber contained in the manual holder and with the plastic protector tube into the oral cavity of the volunteer for 2 min (*see* **Note 10**), as it is shown in Fig. 3a. During the in-mouth extraction, the lips should be kept closed around the plastic tube containing the SPME fiber, and swallowing should be avoided (*see* **Note 11**).

After 2 min of extraction, remove the fiber from the mouth and swallow once.

Remove the fiber from the manual holder and keep it in the freeze (4 °C) until its analysis by GC-MS. For that, introduce the SPME fiber in a sealed glass test tube to assure the proper preservation of the extracted aroma compounds from the breath (*see* **Note 12**).

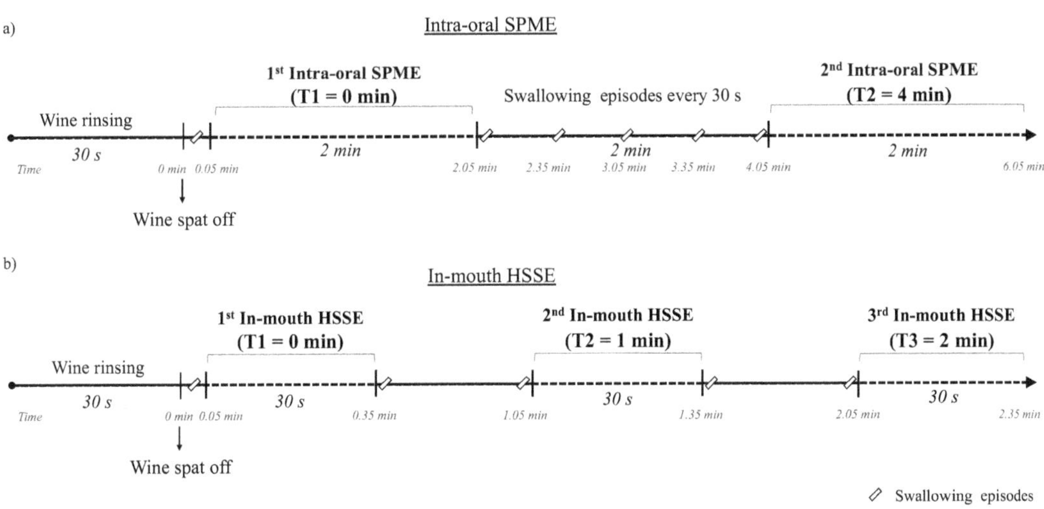

Fig. 2 Scheme of the procedures followed for the aroma extraction in the oral cavity. (**a**) The intra-oral SPME; (**b**) the in-mouth HSSE. (Modified from Perez-Jiménez et al. [4, 5])

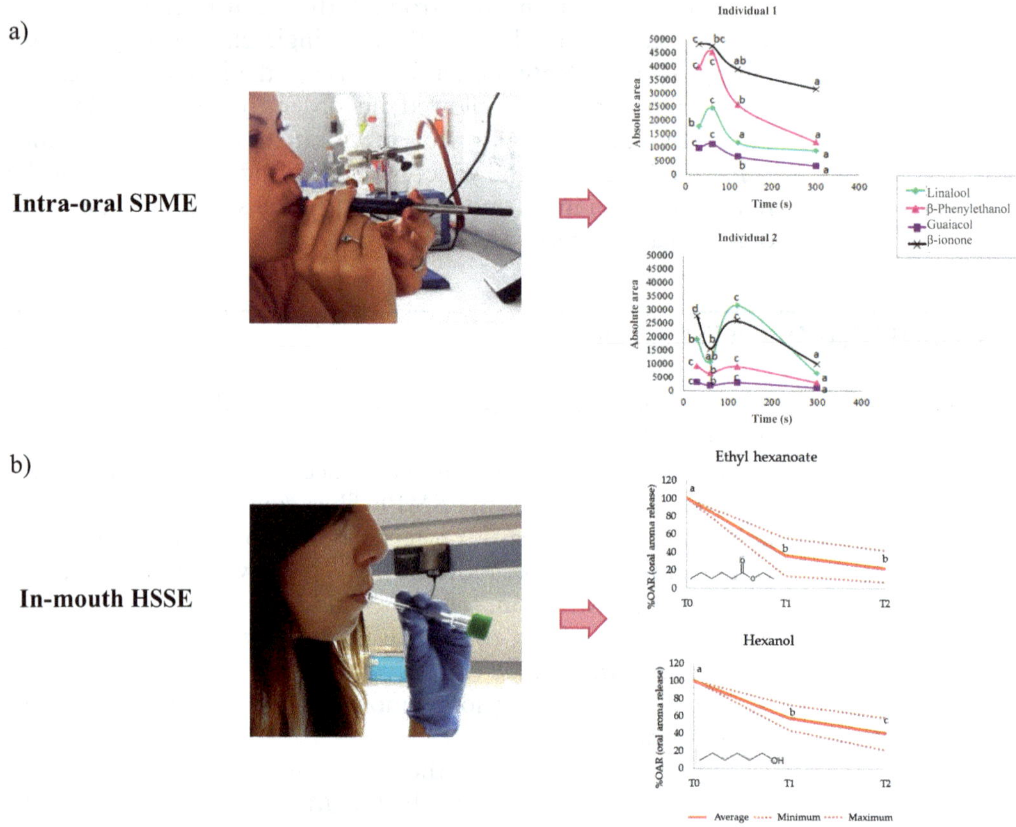

Fig. 3 Picture of the aroma release kinetics of different types of wine aroma compounds obtained by using intra-oral SPME (3a) and in-mouth HSSE procedures (2b). (From Pérez-Jiménez [17])

After the first monitoring, wait 2 min until the second expectoration (*see* **Note 13**). During waiting, the volunteer must breathe normally through the nose and with the mouth closed (*see* **Note 14**) and perform one swallowing every 30 s (five times in total) (Fig. 1a).

Then, 4 min after wine expectoration, perform the second intra-oral monitoring by using a second SPME fiber and following the same instructions than for the first monitorization. After 2 min of intra-oral extraction, remove the fiber from the mouth and desorbed it immediately in the GC system for the analysis of odorants. When the run has finished, analyze the fiber with the breath aroma extract corresponding to the first monitorization.

5.2 In-Mouth HSSE

A schematic representation of the in-mouth HSSE procedure is shown in Fig. 2b.

Expose the glass device containing the twister to the headspace of the mouth and keep the lips closed and placed on the surface of the glass notch (*see* **Note 10**). During the in-mouth extraction swallowing is not allowed (*see* **Note 11**). Keep this position during

30 s as it is shown in Fig. 3b, then, remove the device from the mouth and swallowing once. After 1 min of waiting (*see* **Notes 13** and **14**), swallow once and place into the mouth the second glass device with the second twister inside for the second in-mouth aroma extraction. Keep this position during 30 s, then remove the device from the mouth, swallow once and wait during 1 min until the third breath monitoring. For the third (and followings) in-mouth aroma extractions, repeat the procedure explained for the second monitoring.

Once the in-mouth extractions are finished, remove the twisters from the glass devices using a magnetic bar (*see* **Note 15**), dry them softly with a tissue and place them in their correspondent glass tubes in a refrigerated tray for their thermal desorption.

6 Desorption and Analysis of Aroma Compounds

6.1 Desorption and Chromatographic Analysis

The desorption of the fibers must be in splitless mode for 1.5 min at 270 °C in the injector port of the GC (*see* **Note 16**). The Helium flow is usually set at 1 mL/min. Different chromatographic conditions can be used. For instance, the oven temperature ramp can be start at 40 °C for 2 min, then increased at 8 °C/min to 240 °C and hold it for 15 min.

The desorption of twisters in the TDU is in splitless mode, and the ramp temperature could be start at 40 °C, then increases to 240 °C at 60 °C/min and hold for 5 min (*see* **Note 17**). For the cryofocusing in CIS by using liquid nitrogen the ramp temperature can be start at −100 °C, then heated to 240 °C at 12 °C/min, and hold for 5 min. The injection is usually configured in solvent vent mode. The chromatographic conditions can be similar to those described for the intra-oral SPME.

6.2 MS Identification of Aroma Compounds

Different conditions of MS can be used. An example can be: transfer line at 270 °C, quadrupole at 150 °C and ion source at 230 °C. Electron impact mass spectra is usually recorded at 70 eV and the ionization current is 10 μA. For the acquisitions, both selected ion mass monitoring (SIM) and full scan mode (mass range of 35–350 m/z) can be used depending on the objective of the study.

For the identification of compounds, the mass spectra and retention times are compared with those present in MS libraries (e.g., NIST 2.0 database).

In the intra-oral SPME and in-mouth HSSE methods, the absolute peak areas of aroma compounds are obtained which are used to express the amount of aroma release. These procedures allow to compare the extent of intra-oral aroma release among wine samples by comparing the absolute peaks areas of the same aroma compounds among samples, individuals, etc. Figure 3 shows an example of the release kinetic that can be obtained with the intra-oral SPME and the in-mouth HSSE procedures.

7 Notes

1. For studies aiming to investigate the effect of wine matrix, it would be recommended to recruit at least 8 volunteers. While, for studies aiming to investigate inter-individual differences, a larger number of volunteers would be recommended for their recruitment, although this number should not be too high either (e.g., more than 40). For instance, for the intra-oral SPME it would be recommended to recruit up to 20 volunteers (more than 20 participants will extend too much the duration of the study), whereas for the in-mouth HSSE a large number of volunteers (e.g., 30) could be recruited without extending too much the duration of the study.
2. When working with wines that require to reinforce the aroma profile, it is recommended to use a wine with a low aromatic profile for a better detection of the target aroma compounds added to the wine.
3. To cover the glass wine, a piece of aluminum foil or a Petri dish lid could be used.
4. Before the experiments, it is recommended to check if the selected aroma concentrations (in case of spiked or synthetic wines) provide enough sensitivity. It is important to keep the in mind that differences in the physicochemical characteristics of the aroma molecules (e.g., volatility, polarity) can affect the affinity for the PDMS polymer and the chromatographic response.
5. Having a small amount of ethanol at the bottom of the flask helps the aroma compound to dissolve better.
6. Rinses with bicarbonate can be optional, while rinses with pectin solution and water are highly recommended to clean the palate when tasting wines with a high concentration of polyphenols. It is recommended to perform rinses during around 30 s with each of the palate cleansers (bicarbonate, pectin solution, and water). After rinses with all the palate clean solutions, it is recommended to wait 15 min until starting the next assay (in-mouth extraction) with wine.
7. It is recommended to check and select all the SPME fibers and twisters that will be used through the study, considering their similarity in volatile recovery rates, bearing in mind that differences between them should not exceed 5%.
8. A maximum of three wine samples per day, including sampling replicates (e.g. a total of 9 wine samples), and per volunteer is recommended to avoid participant fatigue. Although this number may vary depending on the ethanol/polyphenol content of the wine samples used.

9. To make this task easier for the volunteers, it is recommended that the person in charge of the experiment give the precise instructions of each step of the extraction procedure.
10. It is important to place the fiber or the glass device containing the twister in the headspace of the mouth avoiding any contact with oral surfaces. During the in-mouth aroma extraction, it is also important to adopt a comfortable posture, for example sitting with both elbows on the table and holding the fiber holder or the glass device with both hands. Volunteers must be also informed to breathe through the nose, not to blow through the mouth, not to swallow and not to make abrupt movements during the intra-oral monitoring. It is recommendable trying to follow always a similar protocol of movements and breathing in all the repetitions of the test in order to obtain a good repeatability (less than 15% of variation among replicates). For training the volunteers in the in-mouth extraction techniques, 2 or 3 additional sessions may be necessary.
11. It may be possible the generation and accumulation of saliva in the mouth during the in-mouth aroma extraction. In this case, gently blot the remaining saliva with a paper tissue, trying to not alter the position and without open the mouth.
12. Preliminary experiments have been performed in order to ensure that there were no significant losses of aroma during the storage of the fiber, which was not more than 1 h.
13. The waiting time between the in-mouth monitorization can be modified depending on the aims of the study.
14. During this waiting time between the first and second intra-oral aroma extraction do not talk or open the mouth. Swallowing is only possible when indicated and it is recommended to be quite and keep in the same place and position moving as little as possible (e.g., do not stand up).
15. For the cleaning of the glass devices after in-mouth extractions, a solution of ethanol in water at 60% is used. For that, all the glass devices and the removable glass bar from inside can be immersed in an ethanolic solution for a few minutes and then dry with clean paper.
16. For cleaning the fibers after each injection avoiding any memory effect, the fibers can be placed in the injector of the GC at 270 °C during 10 min.
17. For cleaning the twisters after the in-mouth aroma extractions avoiding possible residual aroma compounds a GC method is used. For this, the TDU can be configured in split mode at 240 °C for 10 min and the CIS temperature ramp set from 180 to 240 °C. The oven temperature can be start in 50 °C during 2 min, increases up to 240 °C and hold for 15 min. This cleaning procedure also allow the cleaning of the glass liners in which the twisters are placed for their desorption.

Acknowledgments

The authors acknowledge all the volunteers that have participated in the different studies in which both methodologies have been developed. The financial support provided by AEI and MICIN through the Project PID2019-11734-RBI00 is greatly appreciated.

References

1. Esteban-Fernández A, Rocha-Alcubilla N, Muñoz-González C, Moreno-Arribas MV, Pozo-Bayón MÁ (2016) Intra-oral adsorption and release of aroma compounds following in-mouth wine exposure. Food Chem 205: 280–288
2. Perez-Jimenez M, Pozo-Bayon MA (2019) Development of an in-mouth headspace sorptive extraction method (Hsse) for oral aroma monitoring and application to wines of different chemical composition. Food Res Int 121: 97–107
3. Buettner A, Beer A, Hannig C, Settles M (2001) Observation of the swallowing process by application of videofluoroscopy and real-time magnetic resonance imaging—consequences for Retronasal aroma stimulation. Chem Senses 26(9):1211–1219
4. Perez-Jiménez M, Chaya C, Pozo-Bayón MÁ (2019) Individual differences and effect of phenolic compounds in the immediate and prolonged in-mouth aroma release and retronasal aroma intensity during wine tasting. Food Chem 285:147–155
5. Pérez-Jiménez M, Muñoz-González C, Pozo-Bayón MA (2021) Oral release behavior of wine aroma compounds by using in-mouth headspace sorptive extraction (Hsse) method. Foods 10(2):415
6. Muñoz-González C, Criado C, Pérez-Jiménez M, Pozo-Bayón MÁ (2021) Evaluation of the effect of a grape seed tannin extract on wine Ester release and perception using in vitro and in vivo instrumental and sensory approaches. Foods 10(1):93
7. Lyu J, Jianhua F, Chen S, Yan X, Nie Y, Tang K (2022) Impact of tannins on intraoral aroma release and retronasal perception, including detection thresholds and temporal perception by taste, in model wines. Food Chem 375: 131890
8. Pérez-Jiménez M, Muñoz-González C, Chaya C, Fernández-Ruiz V, Álvarez MD, Herranz B, Pozo-Bayón MÁ (2022) Insights on the effect of age and gender on in-mouth volatile release during wine tasting. Food Res Int 155:111100
9. Pérez-Jiménez M, Rocha-Alcubilla N, Pozo-Bayón MÁ (2018) Effect of saliva esterase activity on Ester solutions and possible consequences for the in-mouth Ester release during wine intake. J Texture Stud 50(1):62–70
10. Buettner A (2004) Investigation of potent odorants and afterodor development in two chardonnay wines using the buccal odor screening system (BOSS). J Agric Food Chem 52(8):2339–2346
11. Buettner A, Welle F (2004) Intra-oral detection of potent odorants using a modified stir-bar sorptive extraction system in combination with Hrgc–O, known as the buccal odour screening system (BOSS). Flavour Fragr J 19(6):505–514
12. Yu Y, Chen S, Nie Y, Yan X (2022) Optimization of an intra-oral solid-phase microextraction (Spme) combined with comprehensive two-dimensional gas chromatography–time-of-flight mass spectrometry (Gc × gc-Tofms) method for oral aroma compounds monitoring of baijiu. Food Chem 385:132502
13. Esteban-Fernández A, Muñoz-González C, Jiménez-Girón A, Pérez-Jiménez M, Pozo-Bayón MA (2018) Aroma release in the oral cavity after wine intake is influenced by wine matrix composition. Food Chem 243:125–133
14. Pérez-Jiménez M, Rocha-Alcubilla N, Pozo-Bayón MÁ (2019) Effect of saliva esterase activity on Ester solutions and possible consequences for the in-mouth Ester release during wine intake. J Texture Stud 50(1):62–70
15. Muñoz-González C, Pérez-Jiménez M, Criado C, Pozo-Bayón MÁ (2019) Effects of ethanol concentration on oral aroma release after wine consumption. Molecules 24(18): 3253
16. Muñoz-González C, Pérez-Jiménez M, Pozo-Bayón MÁ (2020) Oral persistence of esters is affected by wine matrix composition. Food Res Int 135:109286
17. Pérez-Jiménez M (2022) Behaviour of aroma compounds during the oral processing of wine: modulation by wine composition and human physiology (Doctoral thesis dissertation)

Part II

Sensory Procedures

Chapter 11

Wine Descriptive Sensory Profiling

Attila Gere and Zoltán Kókai

Abstract

Descriptive methods are an essential part of sensory testing protocols. Out of the selection and training of sensory assessors, testing environment and data analyses have also a significant impact on the obtained results. From a methodological point of view, the selection of an appropriate list of sensory terms might be challenging for researchers. This chapter focuses on the key steps in selecting, training, and maintaining a wine sensory panel as well as gives an overview of the most critical influencing factors (e.g., noise, perfumes, colors, lighting, ventilation, etc.), and provides tips on how to avoid them/decrease their effects. A detailed step-by-step guide will help the reader to conduct successful descriptive analyses. A short guidance is also provided on the most important data analysis steps from analysis of variance (ANOVA) to multivariate analysis methods, such as principal component analysis (PCA) and linear discriminant analysis (LDA). At the end of the chapter, a list of selected papers is presented as examples and inspiration for future researchers.

Key words White wine, Red wine, Trained panel, QDA, Sensory panel

1 Introduction

1.1 Descriptive Sensory Techniques

Descriptive methods are an essential part of sensory testing protocols [1]. In the wine sector, it is especially true that the sensory character of a sample strongly defines the perceived value of that item [2]. There are several possible approaches of performing a descriptive sensory test. The choice of the exact method depends on the goal of the test, the wine samples under analysis, and the mindset of the wine professionals who are responsible for the project.

1.1.1 Overview of Descriptive Methods

The descriptive sensory methods were adapted from the fragrance industry. In that sector, there was a strong tradition of characterizing the samples and differentiating between them based on their sensory attributes. In order to perform that a specialized sensory

Authors Attila Gere and Zoltán Kókai have equally contributed to this chapter

María Ángeles Pozo-Bayón and Carolina Muñoz González (eds.), *Wine Analysis and Testing Techniques*,
Methods and Protocols in Food Science, https://doi.org/10.1007/978-1-0716-3650-3_11,

vocabulary was developed, accompanied by the measuring of the odor compounds' intensities. With the increasing market competition in the food, beverage, and wine sectors, stakeholders realized that through the use of descriptive sensory tests, it is possible to express and communicate the quality of their products [3]. The main criteria of implementing those tests were the selection, training, and maintenance of the panel. The first widely accepted method was called the Flavor Profile (FP) method [4]. During that test, panelists created the list of sensory attributes of the samples' flavor under the supervision of the panel leader. This was group work (consensus), and they evaluated the samples together, creating a flavor profile based on single-intensity values. The next major development was the improvement of Quantitative Descriptive Analysis (QDA), [5]. In that test, the attribute list was still generated by the panel, but the evaluation was performed by the individuals based on on-line scales. With that approach, we produce a more robust dataset that is easier to analyze with statistical tools. The third milestone was the Spectrum Method, where there are extensive attribute lexicons [6]. Based on this, it is possible to skip the step where panelists collect the attributes themselves. Instead of that, they can import them from these databases.

If we look at a wide range of standards and industry-specific good practices, there are currently at least three major fields of application of descriptive test methods.

The first is an approach where the sensory quality of a sample is summarized in one numerical value. This technique is especially useful when the test items have to be categorized in quality grades or medals are assigned in competitions. Although there are several types of that protocol, there is a common feature in all of them: the final result is a number, and usually the higher number means higher quality. This number is composed from a number of sensory attributes (in the case of the simplest systems this is 4 or 5, while more sophisticated ones can go over 10). For trade commodity grading or wine competition sessions, this approach is suitable, but it is not the best choice for research studies.

In the second class of test methods, the goal is to characterize and compare the samples to each other. By this way, we get a better understanding of similarities and differences among the samples. This technique is especially suitable for research projects, or for collecting those attributes which might be used for product attribute communications. The output of that test is a sensory profile, which describes the samples and serves as a basis for comparing other wine items to it. Currently, in most research papers we find that approach as descriptive analysis.

The third application is focusing on quality assurance. In this case, a relatively short list of key attributes is checked. Panelists measure the intensities of these attributes and compare them to the ideal or expected target values. This method provides valuable feedback for winemakers to identify those sensory characteristics which are out of the pre-defined range. Based on their expertise, they can find the root causes of these deviations.

1.1.2 Descriptive Analysis

Descriptive analysis of wine samples requires the skill from the panelists to break down the complexity of the wine to a wide range of single sensory attributes. The nature and the number of those attributes may vary from one test to the other, depending on the goal of the test and the nature of the samples.

One of the most approved descriptive methods in wine sensory research is the profile analysis [7]. This method provides an in-depth analysis of the samples, through the number of attributes selected and also their recorded value These attributes ideally have to be one-dimensional, in other words, they have to describe only one factor in the sensory matrix. An example might be *acidity*, which is measured from *not perceptible* to *very intense*. The measurement of those attributes is performed either on a scale or with scores. In both cases, it is essential to understand that the values that assessors give to the samples measure intensity and not quality. This method is also called Quantitative Descriptive Analysis (QDA) in many papers. The technique relies on the interactive and successive work of the sensory panel.

Descriptive methods also involve scoring systems, which might be tailor-made versions of local wineries or wine competitions. There are also internationally accepted systems, such as the International Organization of Vine and Wine [8, 9]. In that scoresheet, wines can collect a maximum of 100 points. This facilitates the easy expression of results since the scores are in percentage form. At the different categories of attributes, there are different scoring ranges (limpidity has a maximum value of 5, while taste quality can be as high as 14). In the official OIV document, there is a guideline for each attribute and score value. This type of descriptive analysis (scoring) is summarized in a total score at the end of the test. If we look on the sub-scores of the items, we might have further information, but the basic goal of this technique is to assign the samples to quality classes or medal grades. There is a good example from a German wine competition [10], where the sensory profile of the wines is also communicated along with their medal categories, details are shown in Fig. 1.

There are also alternative, promising descriptive techniques, such as free choice profiling, napping, or temporal dominance of sensations [11]. These are usually applied by researchers in academic projects. Due to the development of sensory science in the food and beverage sector, the number of available techniques will probably increase in the future for the wine sector's stakeholders.

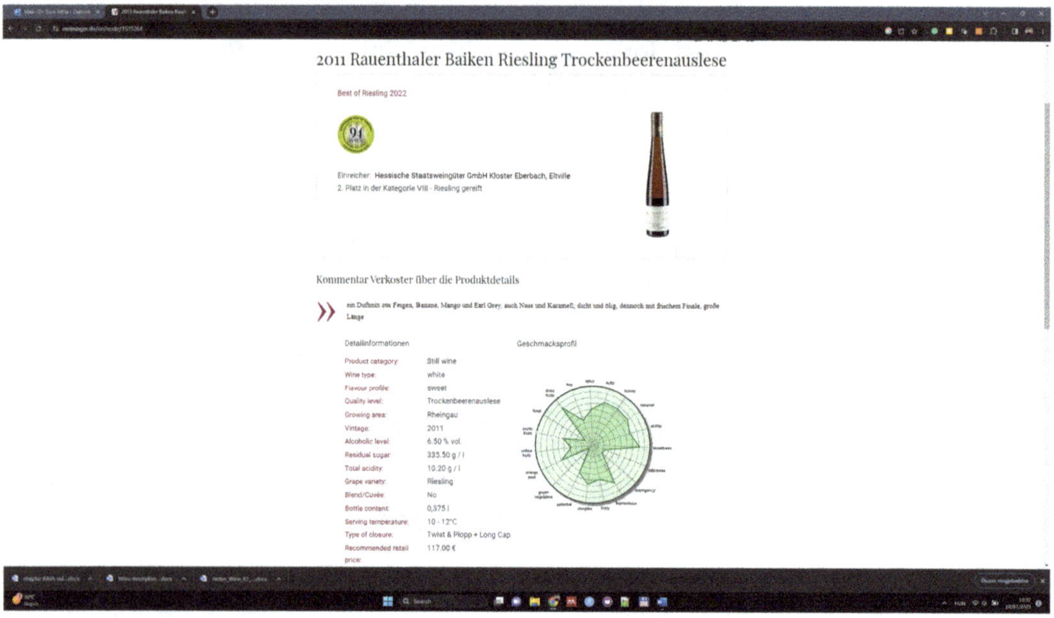

Fig. 1 Example of the results from Mundus Vini wine competition, where sensory profile and detailed description of the wine are also presented along with its medal category and score [10]

2 Methods

2.1 Sensory Panel

The sensory panel is the measuring instrument of the sensory laboratory [12]. Its establishment and maintenance should be considered as an investment of time, human resources, know-how and financial resources. The quality of the panel defines the quality of the data they provide for business decision-makers, winemakers, and QA managers. It should be clear for all involved partners, that the variance of sensory data is always higher than the variance of instrumental data. However, it does not mean that we have to allow very high variation or neglect outlier data. There are several methods to measure and evaluate the performance of a descriptive sensory panel in order to improve the quality of their work.

2.1.1 Selection

Panelist selection is the first step of performing any descriptive analysis. It is desirable that the panel supervisor should select the possible candidates on several criteria. In the wine sector tasting of wine samples was always the part of quality evaluation. If any company or organization decides to implement that activity on a more formal and regulated way, we have to understand that standards [13] and best practices are available on that field. The selection of panelists should rely on their sensitivity, willingness, experience, and motivation. There are wine experts who have considerable experience in practical testing for years or even decades. If they are also integrated in the selection and training process, we have to communicate, that we do not underestimate their knowledge and skills, however, the standardized methodologies require that all panelists need to have a documented history of selection and training. An important point here is, that the selection tests should never focus on the possible weak point of the panelists. Each panelist has strong points, and skills we can use in the panel work. The panel supervisor needs to be aware that people might have negative feelings toward tests, so it should be clear from the first phase, that we look for skills and want to develop those.

The first phase is the screening for those individuals who can be involved in the panel selection procedure. In the wine sector, the internal panels are the most frequent ones. It means that all panel members are the employees or owners of the organization. This technique has the strong advantage that panelists are available in most of the time. On the other hand, it is less likely that they would report a weaker sample or a sensory fault because they are dedicated to the goodwill of their organization. We have to communicate with them, that they should be honest and straightforward even if they have negative impressions with some of the samples. These feedbacks help the winemakers to identify the weak points of the technology and improve them. Theoretically, we can also utilize an external panel. In that case, panelists are not full-time employees of the organization; they are contracted for the part time activity of testing. In some areas of food and beverage industry, we find examples of this panel. The advantage of that technique is that panelists are more neutral toward the tested samples (do not link so strongly to the organization). In this case, we have to consider the costs of panel working hour's fees and the accessibility of the testing site. External panels usually work in towns and cities, but not really suitable for rural areas. The third type is the mixed panel when there are not enough panelists at the organization, so we hire additional candidates.

When the list of candidates is complete, we can organize the selection sessions. Before participating any tests, it is necessary to provide all individuals with an informed consent form and a declaration of allergies and other risks. In that document we have to clarify that panelists understand and take the risks of wine testing, they have no allergic reactions or do not suffer from pollen or other allergies and they can handle the risk of alcohol consumption. All forms have to be recorded and stored according to the local General Data Protection Regulation (GDPR) rules. It is also necessary to generate an individual identification (ID) for all candidates. The recording of their selection and training data will be linked to that individual ID. With this coding, the total results of the group can be viewed by any participants, since only the individual performance data will be identified by the viewer.

Selection tests are available in many forms. Most frequently we rely on the ISO or OIV guidelines [8, 14]. These protocols give a detailed description of color vision tests, odor identification tests, taste threshold, and taste recognition tests. These methods are called as general selection tests since they focus on basic sensory stimuli. We can modify them according to our needs and specialties (*see* **Note 1**).

Selection tests should start with practicing those protocols. We have to provide our panelists at least one occasion, when they can understand and perform the test without analyzing their results. This way they gain a routine with the protocol, and later they can focus on the perceived stimuli. When the practicing period is over, we can organize the selection sessions. Not all tests should be on the same day since that could cause sensory fatigue. Another important factor is, that ideally, we should provide at least two occasions for each test. One testing day might have a bias from being tired or any health issues.

The final step of selection is the evaluation of the data. It is recommended to tabulate all data first, and then analyzing these and looking on trends and patterns. If we set up a pre-determined test criteria before the tests, there is a risk, that either too few or too many candidates will qualify after the testing. The best practice is possible arranging candidates in a rank, and choose those from the first part of the rank, who are sufficient in number for testing. Please notice that in the following phase (training and actual testing) some panelists will disappear. Also, not all panelists will be available at a single descriptive test session. Some documents recommend, that if we need 10 panelists for a single test session, then 20 panelists should be selected from 40 candidates. These numbers are relatively high for small- and middle-scale companies, but the ratio of these numbers carries an important message. There is a decrease of candidates at the selection phase and another at the training and testing phase (*see* **Note 2**).

2.1.2 Training

Training of wine descriptive panelists have to be product specific. We welcome anyone who have experience in any other field of testing, but each product has a special language that the panel have to learn and understand. In the first part of training, it is advisable to use reference materials or reference standards. These are usually soluble chemicals, which are available commercially in kits. Some of them are aromas, that panelists can smell, and some of them are flavors, which can be diluted to water or base wine. Some wine experts reject the use of those kits, since according to their opinion these standards are too artificial, and do not give the complex impression of a complete wine sample. We have to clarify that those training kits do not substitute for the testing and analysis of wine samples. These kits prepare the panelists for the identification of odor and flavor attributes. The following list is just an example of available training kits at the time of writing this chapter.

- *Le Nez du Vin:* probably the earliest example of providing product specific smelling kits for wine experts and wine lovers. Aromas are packed in glass vials, during the training panelists can smell and identify them. After the opening of the vial, we should start with the sniffing of the cap (here the aroma is more diluted). This is followed by the smelling of the glass vial. We have to re-cap the vials as soon as possible to minimize the spreading of the odor and the sensory fatigue. Currently, there are several kits available (Masterkit, red wine, white wine, oak, and faults). Another strength of these kits is that they are available in several languages. Kits contain instructions, aroma description cards, and booklets. In some kits, there is a detailed list of those wines, which strongly carry a given aroma quality [15].
- *Aroma Academy:* they provide wine kits and spirits kits. In this case, the small vials can be opened, and we can use smelling strips, which dilute the odor intensity. Wine 21 odors, Wine 24 odors, and Faults kits are available [16].
- *La symphonie du vin:* contains 32 vials with product-specific aromas. Smelling strips and a black wine-tasting glass is also a part of the package. The availability of this kit is unclear [17].
- *Aroxa:* they provide soluble flavor kits. The contents of the capsules should be dissolved either in water or in a neutral base wine. The panelists can smell and taste these standards. The items can be purchased as individual flavors or in kits, such as Wine Essentials kit, Wine off-flavors kit, Wine Taints kit, and Wine Uno kit. The Uno kit is an idea, when in a single box all capsules contain a different flavor quality [18].
- *FlavorActiv:* the reference standards are soluble compounds. Currently, there are 33 different flavor qualities. Capsules are packed in blisters in accordance with Good Manufacturing Practice (GMP) principles [19].

It is also possible to make the reference standards in our own sensory laboratory. In that case, neutral base wines should be selected first. After that, we have to add fruits, spices or other ingredients to the base wines for a short time (maximum 1 h) soaking. After filtering the items, we should label them with 3-digit random codes. First, give your panelist the base wine as it is in a dedicated glass (labeled as "base"). Then give a series of different references, and instruct them to describe their impressions. We can decide whether to give them a list of possible flavors or aromas, or just a blank paper. At the beginning of the training phase, a list of aromas is a better solution, to give them ideas to choose from. As they proceed with the training, they should work with the blank paper version also. An advanced level is the case when we mix more than one aroma in the sample. The assessors have to identify all the aroma compounds which are perceptible in the wine sample.

During the training, we also have to give specific knowledge on the method of descriptive analysis tests. In case of profile analysis, panelists have to understand that the attributes are measured on intensity scales, so the higher intensity value has no relationship with the quality of the wine. When we train a panel for wine competition, or for quality grading, we have to emphasize the possible product faults and also the scoring systems main categories and the distinction between them (*see* **Note 3**).

2.1.3 Panel Performance Monitoring

Since the sensory panel is the measuring instrument of the laboratory, we need to calibrate them and understand the similarities and differences of the panel members. There is a relevant ISO standard [20] which shows the possible tools and methods to analyze the data of a descriptive panel. This involves a large number of statistical methods, which might be complicated for many sensory practitioners. In order to support the work of panel supervisors, a specialized software was developed, which is freely available at the current time. The software is called PanelCheck (https://www.panelcheck.com/), and it uses a large array of graphical representation of statistical outputs. With the application of these methods, we can identify those attributes, where the panel shows a poor level of consensus. In this case, we have to re-train in those attributes, with such samples, where panelists can easily perceive and understand the actual perception linked to the name of the attributes. If the re-training is not effective, we might consider to remove those sensory properties from the evaluation system. Another useful feature of the software is that we can detect those panelists who are giving outlier values during testing. This can be originated from differences in perception, the different use of evaluation scales, or some misunderstanding of that feature. More relevant case studies in that field can be found in the publications of the PanelCheck Developers [21].

Each time a panel performance analysis is performed, it has to be implemented and communicated in a fair way. No panelist should feel itself inferior, because of some deviation in the performance. The goal of that analysis is to help the harmony among panel members, and it is a longer procedure. All results of such tests should be blind-coded, each assessor has an individual ID number, so the general data protection rules are followed.

Special care should be also paid in chosen the adequate sensory descriptors (*see* **Note 4**).

2.2 Testing Environment

The actual testing environment may depend on several factors. Sensory testing is an integrated part of the whole winemaking process, so there are *less formal tasting sessions* when the oenologist checks the quality of a given batch in order to decide upon the necessary next steps in the technology. In these cases, the environment is usually the winemaking plant. During such tests, we have to avoid any biasing factor, such as scented hand detergents or any other odoriferous substance (e.g., chemicals) in the area. The person or people involved in the testing should focus on the key quality parameters of the wine; during the test, no conversation is allowed. The impressions on the given batch should be recorded in written form. After the test, the participants can discuss their viewpoints with each other.

If a more accurate test is required, the sample is usually taken to the quality assurance laboratory of the company. Depending on the size of the organization this can be an own laboratory, or a contracted, third-party company. Even in smaller laboratories, one can organize a testing protocol, which fits the principles of the good sensory practices. In the *laboratory setting*, if no standardized sensory booth is available, panelists have to be separated by sufficient seating distance. During the test, speech and other activities in the lab should be avoided. In this testing environment, it is easier to find the ideal serving temperature of the samples. Sample preparation should take place in a different room, or if it is not possible, in the same room before participants arrive. Panelists should record their data either by paper and pencil or through an electronic device (laptop, tablet, or smartphone). After the test session, participants are usually encouraged to share their impressions and recorded data with each other.

When the sensory test is performed in a *standardized sensory laboratory* it is usually established according to ISO 8589 [22] or similar documents. The common point in this system is the use of testing booths, which are separated working areas for the panelists. There are different ways of implementation, the booths can be fixed or mobile, and also vary in their physical dimensions and colors. We have to be aware that participants should feel themselves comfortable in these booths, so it is not recommended to build small units in order to increase their numbers. Sensory booths are not always in use, so there are alternative solutions, when the furniture can be transformed into a meeting room. These specialized installations however are rather costly.

Wine competitions has a strong tradition how they seat the panelists and organize their test sessions. In many cases, participants are seated around large tables, so there is no physical separation among them, but the seating distance is relatively large. This setup provides easy access for technicians who serve the samples during the event. Photos and videos can also be taken, without disturbing the panelists' work and in compliance with the GDPR regulations (*see* **Note 5**).

2.2.1 Influencing Factors (Perfume, Colors)

During the sensory tests all those factors should be minimized, which can distract either the attention of the panelists or their sensory impressions. Participants have to be aware that these restrictions are necessary in order to get reliable and reproducible data.

Colors can influence both the evaluation of the wine samples' appearance and also their sensory perception. For the testing area, it is recommended to use white or matte gray surfaces. Walls of strong colors or colored furniture can distract and also can interfere with the appearance of the wine. In dedicated sensory laboratories, panelists often wear white gowns. That is not always feasible in wine sensory testing, but we can inform panelists to avoid clothing of strong or dominant colors.

Personal cosmetics can also interfere with the odor and flavor perception. On the day of testing, panelists should be asked to avoid the use of perfumes or aftershaves. We also have to consider that these aspects are varying in different cultures. Wine testing experts can understand that the neglect of such items on testing days do not mean that they are not in their best condition. Scented hand creams, scented hand sanitizers, and lipsticks should be also avoided. Since hygiene is an important issue, we have to provide panelists with sufficient hand soaps in unscented versions. These are usually available in such products which are designed for people with sensitive skin (hypoallergic products), or specially for laboratory use. In the testing room, we should not use any additional fragrance. The washing and rinsing of glasses should be performed with such detergents which are free from odors.

When someone organizes a wine testing session in a winery or at a conference event, there are additional factors, which might distract the attention of the panelists. In case of a winery, we should be careful to perform it at such a location, which is not influenced by any strong odor (either from the technology of wine making, or from the storage of other materials). When a wine test is implemented at a conference event, one should consider, that the catering zone might have strong odors of different food offered for the guests. With careful planning these issues can be handled.

2.2.2 Noise

Sensory testing is a special laboratory measurement, when we use panelists instead of instruments. It is a general trend, that assessors like to express and share their impressions and emotions about the tested items. There are special situations (like guided tasting sessions, or training of panelists) when this kind of communication is desirable. In general, we have to ask our panelists to work individually. If they have questions or would like to give feedback, primarily they should notify the test supervisor. This silent period during the test helps the concentration of the assessors and also show the differences between their perception. In descriptive sensory analysis there is a special phase, when we encourage the panel to openly discuss their impressions. This is the step, when the individual vocabularies have been recorded and panelists are seated around a table with the samples and their attribute list. In that open discussion they will be able to find similarities and differences between the words they used, and other panelists have used. We have to be aware that during this step not the wine samples are under analysis, but the attribute lists of the panel members. The main aim here is to create a set of attributes that are understandable for all panelists and will be used during the tests to characterize the samples. With this dialogue, panel members will better know and understand each other and create a common language for their work.

In addition to human speech, there are also other possible noise factors, which should be minimized. Such noise sources can be an equipment which operates relatively loudly (refrigerator, air conditioner, photocopy machine, coffee machine, doorbell ring, etc.). If the testing place is close to heavy car, truck or forklift traffic that can be also distracting. Finally, the staff during setting up the tests and while the test is running should be silent. Smartphones should be also put in silent mode during testing.

2.2.3 Lighting, Air Ventilation, Temperature, and Humidity

Although the natural daylight would provide the best testing conditions, it varies greatly over time and also at different geographical locations. ISO 11037 [23] recommends alternative techniques, such as the artificial daylight (D65 light source). The light source should give sufficient strength and amount of light over the working area. There is a frequent opinion of electrical fitting experts, that D65 is too 'strong' for a working environment. However, a descriptive test session does not require a whole working day, so we can stick to the use of that light source, since panelists will spend in the testing area only a few hours per day. For special cases we can mask the color differences between the samples with strong, monochromatic red light. There are also some research papers about especially tunable testing booths, but that is usually suitable for university research laboratories.

Air ventilation should be sufficient, but not disturbing the work of the panel. The local laws and standards about indoor quality have to be followed. Air conditioning is an additional feature, which is not compulsory, but can be advantageous in such locations where both temperature and humidity is high. The target temperatures have to be set in compliance with the comfort of participants and also with the sustainable running of the system. The environmental conditions of the room might have some influence on the serving temperature of the wine samples (*see* **Note 6**).

2.3 Testing Protocol

The main steps of setting up a descriptive profile are listed as following:

1. First, the goal of the test is defined, the samples are selected, and the sensory panelists are notified about the test session in due time. Then, the panel leader gives a general explanation of the purpose of the sensory test, the essence of the method and the tasks to be carried out by the panel.
2. The trained sensory assessors receive a set of samples with 3-digit randomly generated sample codes. The samples are served in an ISO wine tasting glass (ISO 3591:1977 Sensory analysis—Apparatus—Wine-tasting glass, [24]) and each glass should be filled with the same temperature and volume of sample. Glasses are placed on a tray with taste neutralizer as well (*see* **Note 7**).
3. Working individually in sensory booths, the panelists create a list of all the sensory characteristics they perceive in any of the samples [25]. This individual vocabulary can be recorded by paper and pencil, but it is also possible to use an online platform for collecting the panelists' responses. For that phase we have to provide sufficient time for the assessors. If we do not have a sensory booth system, seat the participants with suitable separation distance (*see* **Note 8**).
4. Through group work—using the so-called consensus method—the assessors define the characteristics that were clearly understood and perceived by all of them. Definitions for the sensory attributes are also created in this step. Panelists sit down to a large table, with their samples and their individual vocabulary. The panel leader moderates this session. Panelists are asked to share their vocabulary with the group. This can be performed in structured way, for example, first they collect all the attributes related to appearance, then odor characteristics followed by flavor qualities and aftertaste. Each attribute is displayed either on a whiteboard or a smartboard. When all items are displayed, the group starts to discuss the list. All these items should be removed, which are hard to define, not perceived by the majority of the panel or related to personal preference. After that we look for synonyms when different words carry the same meaning. Here it is especially useful, that the samples are available in front of each panelist, so they can re-taste and discuss those items, where there is no consensus yet. At the end of that step, there will be a concise list of those attributes which are perceived an accepted by the whole panel.
5. Using reference samples/materials. Using a reference sample or reference material reduces the standard deviation in the obtained sensory data. Reference samples are usually used

when the aim is to compare other samples to a golden standard. In this situation, the sensory attributes of the reference sample are recorded on the sensory ballot and the reference sample is also available for the assessors. This way, the reference sample gives an additional anchor on the sensory ballot and helps the assessors in their work (*see* **Note 9**). The use of reference materials is more widespread. Reference materials are defined for the sensory attributes to help the assessors in their work. In this situation, not a complete sample but different materials and their corresponding intensities are presented. For example, the reference material for sensory attribute cooked vegetables can be 2 tsp. of canned green beans water [26] (*see* **Note 10**).

6. Based on these, a scoring system (sensory ballot) is created (sensory attributes, scales). The sensory attributes are usually scored using a linear scale, with labels describing the two endpoints. It is also advisable to include space for additional comments at the end of the sensory ballot. There is an alternative way to save time when creating the sensory ballot. Instead of following points 2–4, a predefined list of sensory attributes can also be used. This predefined list can either be a list that was created earlier by the panel, or it can be adapted from other studies as well (*see* **Note 11**). A possible source of such adaptations could be the international literature by finding a well-established paper that uses similar samples and presents how the list of attributes has been defined (*see* **Note 12**).
7. The same samples are then served again but with different 3-digit randomly generated sample codes, that not only differ from the ones used in 2) but in this time they are different among the assessors as well. Not only the codes but the order of the sample should be considered. Randomized presentation of the samples is necessary to avoid the carry-over effect, that is known as a sensory bias caused by the residual sensations of previously served samples [27].
8. The prepared samples are then assessed by each assessor on the sensory ballot. The sensory test can be done on computers (laptops, tables, etc.) or using paper and pencil as well. This section is completed with the recording of the data, which will be submitted for statistical analysis. The advantages and disadvantages of online and paper-based sensory ballots is presented by Table 1.

Another option of the protocol is that assessors receive a complete list of attributes from the panel leader, and they learn and discuss the attributes with group work. In this case, the panel adapts itself to the test ballot. This technique is especially useful if the goal of the test is to perform a session based on a standardized test sheet, so there is no option of changing the attributes. A possible use case

Table 1
Advantages and disadvantages of online and paper-based sensory ballots

Online sensory ballots		Paper-based ballots	
Advantages	**Disadvantages**	**Advantages**	**Disadvantages**
Increased data security	Need of digital tools (computer, laptop, tablet etc.), electricity	Higher independency in terms of testing location	Lower data security (lost/damaged ballots)
No need of digitization of data	Need of stable internet access	Better for one-time-use	Data digitization
Automated checking if all fields have been filled	Higher price	Cheaper	Manual checking if all fields have been filled
Possibility of complex ballots	Deeper knowledge on the software	Easy to create ballots	Less features are available
Connection to other services (online surveys, etc.)	In case of any electrical or software issues, the test stops	No digital tools needed, only paper and pencil are needed	Less sustainable
Digital panel maintenance		Independency from technical issues	
Supports multiple language ballots			
More sustainable			
Built-in data analysis			
Increased transparency			

can be when a key customer requests a special protocol based on their central system. It is also a frequent situation, when some panel members are changed, and the new members must learn the system. For them, we must provide every necessary item for practical learning (description of the attributes samples that show these attributes and reference materials for high intensities).

2.4 Data Analysis

The structure of a descriptive sensory data set should contain the panelists (their codes), the samples and the replicates as factors and the all the sensory attributes evaluated. The number of rows, therefore, heavily depends on the first three.

The first step of the data analysis is to obtain the summary statistics. A summary statistic usually involves the number of observations, missing values, minimum, maximum, range, mean, standard deviation, median, mode, and other (if needed) metrics. These metrics help us to identify and issues regarding the input data. All identified issues (missing values, extreme values etc.) should be addressed before data analysis as most of the used methods are sensitive to such issues as well as these issues might influence the results of the applied statistics.

According to Heymann and colleagues [28], the data analysis should involve the following three steps:

1. Multivariate analysis of variance (MANOVA) to test the significance of the factors and their second interactions.
2. Analysis of variance (ANOVA) of the attributes on the significant factors.
3. Linear discriminant analysis (LDA) and principal component analysis (PCA) to map the relationship between the samples and the sensory attributes.

During MANOVA, the aim is to identify the effect of all the factors (and their second interactions) on the measured sensory attributes. If a factor has a significant ($p < 0.05$) effect in MANOVA, then the individual differences should be analyzed using ANOVA. MANOVA does not analyses the individual attributes but the factors. For example, if MANOVA gives significant effect for the factor sample, then we need to run ANOVA to identify the sensory attributes that are significantly different between the samples. Therefore, an individual ANOVA should be computed on all sensory attributes, one-by-one.

As mentioned before, ANOVA determines the differences between the samples, however, ANOVA gives only that there are at least two samples that are significantly different based on a given sensory attribute. Therefore, one additional statistic, the so-called post hoc test is needed to calculate pairwise differences between the samples for all attributes. Although there is multiple option to choose a post hoc test, Fishers least significant difference (LSD) and Tukey's honestly significant difference (HSD) are the two most wiled used ones. The former is somewhat liberal, while the latter is more conservative [28].

As ANOVA and post hoc tests are not visual tools, a suggested method to visualize the results is the profile plot. The quick visualization tool, the profile plot, provides a perfect possibility to summarize the results of descriptive profiles. The profile (or spiderweb, or radar) plot consists of as many axes as sensory attributes the panelists evaluated. The mean values of the evaluated samples are then registered and connected on the profile plot, therefore creating a unique, fingerprint-like pattern of the samples. This way, visual comparison of samples can be done quickly and easily and knowing the differences between the samples across the sensory attributes, it can be used to identify patterns in the profile. The easiest way to create profile plots is using MS Excel, as it includes a built-in function to create profile plots from the mean attribute values of the samples.

Due to the structure of sensory profiles (e.g., multiple variables are measured on multiple samples by multiple assessors and replicates), a wide range of multivariate data analysis tools can also be employed to uncover patterns in the data set.

The connection between samples and sensory attributes are usually mapped by using principal component analysis (PCA), a dimension reduction method that uses the correlation matrix of the attributes to create uncorrelated principal components (PCs). When running PCA, the number of PCs to keep is one of the key questions. In sensory profiling, PCA is mainly used for visualization purposes, therefore two or three PCs are kept. Detailed information on running PCA in R is provided by [29]. As PCA is primarily a dimension reduction method, it helps the user to present the data set in a convenient format so the user can assess the connections between the samples and the attributes. Scores plots are generated from the principal component scores (e.g., the rows of the input data matrix) and therefore present the position of the samples. Shorter distance between samples means higher sensory similarity, while larger distance means higher sensory differences. Loading plots are generated from PCA loadings (e.g., variables), therefore here the vectors represent the sensory attributes. Again, attributes closer to each other are similarly present in the samples. The scores and loading plots are combined in the so-called BiPlot, which scales these two to the same unit and overlays them on each other. This way, the user can assess the connections between the samples and the attributes visually. If an attribute is close to a sample, this means that this attribute was present with higher scores in the sample. For example, CabR can be characterized by sour aftertaste, sour taste and intense off-flavor, while KadH has intense vanilla and redberry odor (Fig. 2) If the attribute is located far away from the sample, then it was probably not perceived by the assessors in the sample.

Classification methods are also widely used to characterize the samples. Classification methods help us determine if the group of samples can be identified or not. The two families of classification methods are unsupervised and supervised classification. During unsupervised classification, the user has no a priori-determined labels for the cases, for example, we either do not know the exact grouping of the samples or we aim to justify the grouping. Although there are multiple different unsupervised classification methods available, the most widely used technique is agglomerative hierarchical clustering (AHC). For a guide on how to run and validate AHC, see [31]. The result of AHC is a so-called dendrogram, that groups the samples based on their similarities and is used to determine sample groups (clusters). If some samples fall into the same cluster, then are considered similar, while the distance between clusters gives the similarities of the clusters.

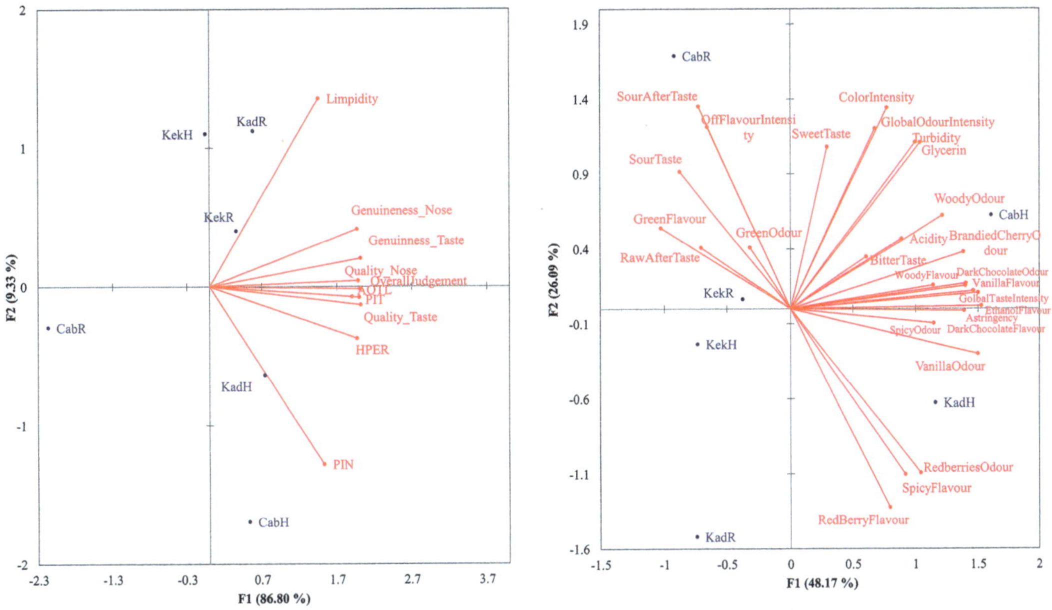

Fig. 2 BiPlot of a principal component analysis presenting the samples (blue) and the sensory attributes (red). While a score plot presents only the samples and the loading plot only the sensory attributes, the BiPlot provides a convenient presentation of both results. *CabH* produced Cabernet franc, *CabR* commercial Cabernet franc, *KadH* produced Kadarka, *KadR* commercial Kadarka, *KekH* produced Kékfrankos, *KekR* commercial Kékfrankos. (Source: Fig. 8 of Guld et al. [30])

When it comes to supervised classification, the labels (e.g., group memberships) of the samples are known by the model. These labels can be determined by the user (e.g., type of wines), or created using cluster analysis. One of the most widely used supervised classification methods in sensory profiling is linear discriminant analysis (LDA), also called canonical variate analysis (CVA). Using LDA, a classification model is built, in which the within-group distances are minimized and the between-group distances are maximized. The model then is able to describe our samples, in terms of defining if the a priori groups can be statistically discriminated, and, if yes, new, unknown cases can be classified into existing classes.

2.5 Examples from the International Literature

There are multiple research papers available in the international literature focusing on wine descriptive analyses. Table 2 provides a selected short list of different studies focusing on the analysis of different wine types and vintages from all around the world. These papers provide a list of descriptive terms as well as reference materials that might serve as a good basis for future wine descriptive studies.

Table 2
List of publications dealing with wine descriptive analysis

References	Wine type	Wine origin	Number of attributes	Number of samples	Vintage	Definition	Reference material
[26]	Cabernet sauvignon	Australia	25	52	2015	Yes	Yes
[32]	Shiraz/Syrah	Australia	34	17	2013–2015	Yes	Yes
[30]	Kadarka, Kékfrankos, and Cabernet franc	Hungary	27	6	Varied	Yes	Yes
[33]	Tannat	Uruguay	9	13	Varied	Yes	Yes
[34]	Rosé	Australia	25	26	2012–2013	Yes	Yes
[35]	Cava	Spain	9	18	2003–2005	Yes	Yes
[36]	Carbernet Sauvignon/Bordeaux blends with Cabernet Sauvignon	USA	34	14 (Carbernet Sauvignon) 10 (Bordeaux blends with Cabernet Sauvignon)	2000–2009	Yes	Yes
[37]	Chardonnay, Pinot Noir and Pinot Meunier	Australia	28	50	Varied	Yes	Yes
[38]	Rosé	Denmark	56	3	Varied	Yes	Yes
[39]	Port wine	Portugal	23	28	Varied	Yes	Yes
[40]	Petit Verdot	Spain	9	5	2015–2019	Yes	Yes
[41]	Cabernet Sauvignon	Australia	52	16	2015	Yes	Yes
[42]	Tempranillo	USA	33	144	After 12 months	Yes	Yes
[43]	Pinot Noir	New Zealand	53	18	2013 and 2016	Yes	Yes

[44]	Cabernet Sauvignon and Chardonnay	Australia	23	25	2013–2015 (Cabernet Sauvignon) and 2015–2016 (Chardonnay)	Yes	Yes
[45]	Shiraz	Australia	14	3	2016–2017	Yes	Yes
[46]	Malbec	Spain	14	4	2013–2016	Yes	Yes
[47]	Different types (Rioja, Ribera del Duero region)	Spain	13	12	2014–2016	No	No
[48]	White wine Chelva	Spain	12	10	2013–2017	Yes	Yes
[49]	Different types	Varies	12	6	Varied	Yes	Yes
[50]	Malbec	Argentina and USA	35	41	2011	Yes	Yes
[51]	Marlborough Sauvignon blanc	New Zealand	11	12	2009	Yes	Yes

3 Notes

1. Taste identification tests.

   In standardized taste identification tests, the fifth basic taste is the umami (monosodium glutamate). Since this taste quality is not relevant in wines, we can replace it with another stimulus, like astringency (e.g., tannic acid). Similarly, the set of odor samples can be harmonized to those key sensory attributes, which are relevant of the wines to be tested.

2. Mindset of test supervisor—wine experts.

   Before organizing and implementing a descriptive sensory test we have to understand, that wine experts have a very special mindset about the quality evaluation of the wine. First, they look on the wine samples as one integrity, which makes sometimes difficult to fully implement the descriptive protocol, when we want to break down samples to several attributes. Because of that wine experts often involve more complex attributes (balance, complexity, true to variety, overall merit, etc.). All these aspects are important, when we consider the perceived quality of wine; however, it is challenging to define these descriptors and analyze them on a scale. A "proof" of these attributes might be at the statistical analysis when we understand the level of variance over panelists. If deviations are moderate, then the panel handles these attributes uniformly. In case of higher variations, a discussion, clarification, and re-definition is necessary for these characteristics.

3. Opinion of experts about commercially available training kits.

   For descriptive analysis, it is necessary to train our panelists. Some people consider that formal training cannot be compared to years or decades of winemaking and wine testing. We agree, that practical wine experience is very much appreciated, however with the change of generations (alpha, Z-gen, etc.) we have to also consider that these groups vary their job position more often. Therefore, in many cases, we need to perform formal, standard-based training for creating a panel of sufficient size. During that work, the use of commercially available reference kits is very valuable. The test supervisor does not have to develop the references individually and there is no chemical risk assessment since the commercial kits went through full approval. It is a popular opinion, that those kits are too *"artificial"* and *"they do not represent wine as a whole."* We would compare these kits to preparations, where medical students can study the parts of the human body without making any harm. Similarly, the wine panelists can learn the technique of identifying and understanding each aroma or flavor compound through those kits as the first step. This should be followed by wine sample testing, where they can understand the relationship of those attributes, when they appear together in a wine. So, these kits do not substitute the testing of wine, they prepare the panelists for it.

4. Mindset of test supervisor—sensory analyst.

   If a sensory analyst is responsible for the test protocol, it is very likely that the work will be performed on the basis of international standards and good testing practices. Sensory experts focus on issues considering test design, test implementation, statistical quality of the data, and the performance of the panel. They usually have a lower level of experience in winemaking, so they require special input or discussion from winemakers, especially during the selection of samples and descriptors. Sensory analysts have to understand that the wine is always more than the addition of the individual sensory attributes, so not all quality characteristics are measurable by descriptive analysis. Their open-minded attitude toward wine experts may lead to successful synergies.

5. Special test locations (historical places)—testing vs visitors.

   In some occasions, the wine descriptive test is performed at a special location. This might be the dedicated tasting room of the winery or a room in a palace in case of a wine competition. It was a practical question during the organization of such an event that should we allow visitors to look in those rooms or should be fully closed. It might be negotiated, that visitors might go until a certain point (e.g., 2 meters from the entrance). They might be not allowed to take photos or videos and the use of flash should be avoided. If we want to communicate in case of a wine competition, that our work is transparent and independent, a certain level of *openness* might help to achieve this in the common opinion. However, the work of panelists should not be disturbed.

6. Air ventilation and off-odors.

   The incoming fresh air should pass through an active carbon filter, to remove any off-odors. The ventilators should be either silent, when working, or should be housed in such a way, that their noise is not noticeable for the panelists and the sample preparation technicians.

7. Taste neutralizers.

   Non-carbonated, neutral mineral water and/or neutral bread sticks serve well as taste neutralizers.

8. Separation of panellists—physical boots vs. distance seating.

   The isolation of sensory panelists during descriptive analysis went through and goes through a change. In the early version of those tests, (Flavor Profile) panelists were seated at a round table with no physical separation. Later, in the case of

QDA, assessors were working in testing booths, with almost full separation. These booths had many features that standards recommended (standardized lighting source, sample serving window, water tap, and a sink). Currently, there are some laboratories, where the design of booths became more open, so the dividers are lower, or in some cases, panelists have a spacious seating, but no physical separation at all. There are two major issues we have to focus on when designing a descriptive test (1) panelists should feel themselves comfortable and (2) they should work undisturbed, with no influence from others. So, a test with no physical dividers is not necessarily inferior compared to standard testing booths, if care is taken on those two factors. The final decision should be made by the test organizer after discussing it with all relevant stakeholders.

9. Reference samples.

   In case of lack of reference materials, a good option would be to choose a reference sample. This sample will be discussed by the panel in details, and they will assign a value for all sensory attributes being evaluated. This way there will be anchors provided for the panel for all sensory attributes. This will help their work during the testing, but this is not compulsory.

10. Reference materials.

    There are several producers that provide reference materials as discussed earlier. However, there are situations where we need reference materials that are not available and we need to create them. In case of lack of time and/or resources, collecting ideas for the list of sensory attributes from published papers could be timesaving. Be aware to get inspiration from papers that went through the double-blind peer review process and are quality works. If doing so, proper citation of the work is a must. Such list is presented in Table 2.

11. Sensory attributes.

    When we do not have time and/or resources to conduct a two-step descriptive profile, collecting ideas for the list of sensory attributes from published papers could be timesaving. Be aware to get inspiration from papers that went through the double-blind peer review process and are quality works. If doing so, proper citation of the work is a must again. Such list is presented in Table 2.

12. Special attributes.

    There are special attributes in wine analysis, which include "harmony," "balance," "roundness," and "typicality." There is a strong tradition in wine tasting to use those expressions to a certain extent. On the other hand, we must acknowledge, that these attributes are hard to define during a product specific training, and their evaluation may vary from one panelist to the other. Therefore, in the first step of panel performance analysis, we might focus on more objective attributes, which are easier to define and train (e.g., raspberry odor, woody aroma, etc.).

Acknowledgments

The authors thank the support of the National Research, Development, and Innovation Office of Hungary (OTKA, contracts No FK 137577). AG thanks the support of the János Bolyai Research Scholarship of the Hungarian Academy of Sciences.

The list of panel training reference standards mentioned in this chapter. This information is given for the convenience of users:

Le Nez Du Vin, https://www.lenez.com/en/home accessed: 03th of July 2023.
Aroma Academy, https://www.aroma-academy.co.uk/ accessed: 03th of July 2023.
Aroxa, https://aroxa.com/ accessed: 03th of July 2023.
FlavorActiv, https://www.flavoractiv.com/ accessed: 03th of July 2023.

References

1. Kemp SE, Ng M, Hollowood T, Hort J (2018) Introduction to descriptive analysis. In: Descriptive analysis in sensory evaluation. Wiley, New York, pp 1–39. https://doi.org/10.1002/9781118991657.ch1
2. Yang J, Lee J (2020) Current research related to wine sensory perception since 2010. Beverages 6(3):47. https://doi.org/10.3390/beverages6030047
3. Varela P, Ares G (2012) Sensory profiling, the blurred line between sensory and consumer science. A review of novel methods for product characterization. Food Res Int 48(2):893–908. https://doi.org/10.1016/j.foodres.2012.06.037
4. Cairncross SE, Sjostrom EB (1997) Descriptive sensory analysis in practice. Food and Nutrition Press
5. Stone H, Sidel JL, Bloomquist J (1997) Quantitative descriptive analysis. In: Descriptive sensory analysis in practice. Food and Nutrition Press, pp 51–63
6. Munoz AM, Civille GV (1998) Universal, product and attribute specific scaling and the development of common lexicons in descriptive analysis. J Sens Stud 13(1):57–75. https://doi.org/10.1111/j.1745-459X.1998.tb00075.x
7. ISO (2016) ISO 13299:2016 Sensory analysis—methodology—general guidance for establishing a sensory profile. ISO, Geneva. [Online]. Available: https://www.iso.org/standard/58042.html
8. OIV (2015) Review document on sensory analysis of wine. [Online]. Available: https://www.oiv.int/public/medias/3307/review-on-sensory-analysis-of-wine.pdf
9. OIV (2021). OIV standard for international wine and spirituous beverages of vitivinicultural origin competitions. [Online]. Available: https://www.oiv.int/public/medias/7895/oiv-patronage-competition-norme-ed-2021.pdf
10. Meininger Mundus Vini Wine Competition (2023). https://www.meininger.de/en/wine/verkostungen/ergebnisse/mundus-vini-spring-tasting-2023
11. Schlich P (2017) Temporal dominance of sensations (TDS): a new deal for temporal sensory analysis. Curr Opin Food Sci 15:38–42. https://doi.org/10.1016/j.cofs.2017.05.003
12. ISO (2012) ISO 8586:2012 sensory analysis—general guidelines for the selection, training and monitoring of selected assessors and expert sensory assessors. ISO, Geneva, p 1–38. [Online]. Available: https://www.iso.org/standard/76667.html
13. ISO (2017) ISO 6658:2017 sensory analysis—methodology—general guidance. ISO, Geneva, p 1–26. [Online]. Available: https://www.iso.org/standard/65519.html

14. ISO (2011) ISO 3972:2011 sensory analysis—methodology—method of investigating sensitivity of taste. ISO, Geneva, p 1–10. [Online]. Available: https://www.iso.org/standard/50110.html
15. Le Nez Du Vin (2023). https://www.lenez.com/en/home
16. Aroma Academy (2023). https://www.aroma-academy.co.uk/
17. La Symphonie Du Vin (2023). https://www.athenaeum.com/article/1940221-la-symphonie-du-vin-n1-recognize-and-understand-the-aromas-of-wines-english-version%2D%2Dpeugeot-saveurs
18. Aroxa (2023). https://aroxa.com/
19. FlavorActiv (2023). https://www.flavoractiv.com/
20. ISO (2012) ISO 11132:2012 sensory analysis—methodology—guidelines for monitoring the performance of a quantitative sensory panel. ISO, Geneva, p 1–22. [Online]. Available: https://www.iso.org/standard/76669.html
21. Næs T, Brockhoff PB, Tomic O (2010) Statistics for sensory and consumer science. Wiley, New York. https://doi.org/10.1002/9780470669181
22. ISO (2007) ISO 8589:2007 sensory analysis—general guidance for the design of test rooms. ISO, Geneva. [Online]. Available: https://www.iso.org/standard/36385.html
23. ISO (2011) ISO 11037:2011 sensory analysis—guidelines for sensory assessment of the colour of products. ISO, Geneva, p 1–18. [Online]. Available: https://www.iso.org/standard/50112.html
24. ISO (1977) ISO 3591:1977 sensory analysis—apparatus—wine-tasting glass. ISO, Geneva. [Online]. Available: https://www.iso.org/standard/9002.html
25. ISO (1994) ISO 11035:1994 sensory analysis—identification and selection of descriptors for establishing a sensory profile by a multidimensional approach. ISO, Geneva. [Online]. Available: https://www.iso.org/standard/19015.html
26. Souza Gonzaga L, Capone DL, Bastian SEP, Danner L, Jeffery DW (2020) Sensory typicity of regional Australian cabernet sauvignon wines according to expert evaluations and descriptive analysis. Food Res Int 138:109760. https://doi.org/10.1016/j.foodres.2020.109760
27. Torrico D, Jirangrat W, Wang J, Chompreeda P, Sriwattana S, Prinyawiwatkul W (2018) Novel modelling approaches to characterize and quantify carryover effects on sensory acceptability. Foods 7(11):186. https://doi.org/10.3390/foods7110186
28. Heymann H, King ES, Hopfer H (2014) Classical descriptive analysis. In: Novel techniques in sensory characterization and consumer profiling. CRC Press, Boca Raton, pp 18–26
29. Abdi H, Beaton B (2015) Principal component and correspondence analyses using R. Springer. [Online]. Available: https://link.springer.com/book/9783319092553
30. Guld Z, Nyitrainé Sárdy D, Gere A, Rácz A (2020) Comparison of sensory evaluation techniques for Hungarian wines. J Chemom 34(4). https://doi.org/10.1002/cem.3219
31. Gere A (2023) Recommendations for validating hierarchical clustering in consumer sensory projects. Curr Res Food Sci 6:100522. https://doi.org/10.1016/j.crfs.2023.100522
32. Pearson W, Schmidtke L, Francis IL, Blackman JW (2020) An investigation of the pivot© profile sensory analysis method using wine experts: comparison with descriptive analysis and results from two expert panels. Food Qual Prefer 83:103858. https://doi.org/10.1016/j.foodqual.2019.103858
33. Varela P, Gambaro A (2006) Sensory descriptive analysis of Urugayan Tannat wine: correlation to quality assessment. J Sens Stud 21(2):203–217. https://doi.org/10.1111/j.1745-459X.2006.00061.x
34. Wang J, Capone DL, Wilkinson KL, Jeffery DW (2016) Chemical and sensory profiles of rosé wines from Australia. Food Chem 196:682–693. https://doi.org/10.1016/j.foodchem.2015.09.111
35. Torrens J, Riu-Aumatell M, Vichi S, López-Tamames E, Buxaderas S (2010) Assessment of volatile and sensory profiles between base and sparkling wines. J Agric Food Chem 58(4):2455–2461. https://doi.org/10.1021/jf9035518
36. Hjelmeland AK, King ES, Ebeler SE, Heymann H (2013) Characterizing the chemical and sensory profiles of United States cabernet sauvignon wines and blends. Am J Enol Vitic 64(2):169–179. https://doi.org/10.5344/ajev.2012.12107
37. Culbert JA, Ristic R, Ovington LA, Saliba AJ, Wilkinson KL (2017) Influence of production method on the sensory profile and consumer acceptance of Australian sparkling white wine styles. Aust J Grape Wine Res 23(2):170–178. https://doi.org/10.1111/ajgw.12277
38. Liu J, Arneborg N, Toldam-Andersen TB, Petersen MA, Bredie WL (2017) Effect of sequential fermentations and grape cultivars on volatile compounds and sensory profiles of

Danish wines. J Sci Food Agric 97(11): 3594–3602. https://doi.org/10.1002/jsfa.8218

39. Vilela A, Monteiro B, Correia E (2015) Sensory profile of port wines: categorical principal component analysis, an approach for sensory data treatment. Ciênc Téc Vitiviníc 30(1):1–8. https://doi.org/10.1051/ctv/20153001001
40. Delgado JA, Sánchez-Palomo E, Osorio Alises M, González Viñas MA (2022) Chemical and sensory aroma typicity of La Mancha Petit Verdot wines. LWT 162:113418. https://doi.org/10.1016/j.lwt.2022.113418
41. Souza Gonzaga L, Bastian SEP, Capone DL, Danner L, Jeffery DW (2022) Consumer perspectives of wine typicity and impact of region information on the sensory perception of cabernet sauvignon wines. Food Res Int 152: 110719. https://doi.org/10.1016/j.foodres.2021.110719
42. Feng Z, Martínez-Lapuente L, Ayestarán B, Guadalupe Z (2023) Volatile and sensory characterization of Tempranillo wines aged in Quercus alba oak barrels of different geographical origins in USA. LWT 173:114328. https://doi.org/10.1016/j.lwt.2022.114328
43. Araujo LD et al (2021) In-mouth attributes driving perceived quality of pinot noir wines: sensory and chemical characterisation. Food Res Int 149:110665. https://doi.org/10.1016/j.foodres.2021.110665
44. Niimi J, Boss PK, Bastian SEP (2018) Sensory profiling and quality assessment of research cabernet sauvignon and chardonnay wines; quality discrimination depends on greater differences in multiple modalities. Food Res Int 106:304–316. https://doi.org/10.1016/j.foodres.2017.12.060
45. Longo R, Blackman JW, Antalick G, Torley PJ, Rogiers SY, Schmidtke LM (2018) Volatile and sensory profiling of Shiraz wine in response to alcohol management: comparison of harvest timing versus technological approaches. Food Res Int 109:561–571. https://doi.org/10.1016/j.foodres.2018.04.057
46. Sánchez-Palomo E, Trujillo M, García Ruiz A, González Viñas MA (2017) Aroma profile of malbec red wines from La Mancha region: chemical and sensory characterization. Food Res Int 100:201–208. https://doi.org/10.1016/j.foodres.2017.06.036
47. Mora M, Dupas de Matos A, Vázquez-Araújo L, Puente V, Hernando J, Chaya C (2021) Exploring young consumers' attitudes and emotions to sensory and physicochemical properties of different red wines. Food Res Int 143:110303. https://doi.org/10.1016/j.foodres.2021.110303
48. Sánchez-Palomo E, Delgado JA, Ferrer MA, Viñas MAG (2019) The aroma of La Mancha Chelva wines: chemical and sensory characterization. Food Res Int 119:135–142. https://doi.org/10.1016/j.foodres.2019.01.049
49. Mora M, Urdaneta E, Chaya C (2018) Emotional response to wine: sensory properties, age and gender as drivers of consumers' preferences. Food Qual Prefer 66:19–28. https://doi.org/10.1016/j.foodqual.2017.12.015
50. King ES et al (2014) Regional sensory and chemical characteristics of Malbec wines from Mendoza and California. Food Chem 143: 256–267. https://doi.org/10.1016/j.foodchem.2013.07.085
51. Parr WV, Schlich P, Theobald JC, Harsch MJ (2013) Association of selected viniviticultural factors with sensory and chemical characteristics of New Zealand sauvignon blanc wines. Food Res Int 53(1):464–475. https://doi.org/10.1016/j.foodres.2013.05.028

Chapter 12

Rapid and Cost-Effective Methods for Wine Sensory Profiling: Napping and Sorting

Matthew McSweeney

Abstract

Descriptive sensory analysis can be time-consuming and expensive; therefore, sensory scientists have begun to use rapid and cost-effective methods. Napping and sorting are two rapid profiling methods. These methods ask panelists to evaluate wine based on their similarities and differences. Both napping and sorting can identify general product descriptions and the sensory properties that describe different varieties and styles of wine. This chapter presents the procedures for the use of the napping and sorting tasks, as well as the commonly used statistical analysis methods.

Key words Napping, Sorting, Projective mapping, Rapid profiling, Rapid analysis, General product description

1 Introduction

Descriptive analysis and trained panels are one of the most extensively used tools in the field of sensory evaluation. Descriptive analysis is considered the gold standard in sensory analysis and allows for very detailed and reproducible results to be identified [1]. But descriptive analysis is a very time-consuming and expensive process [2]. As such, rapid sensory methods or rapid profiling methods have been created to evaluate products. Rapid profiling methods are able to assess a product in a much quicker time frame than descriptive analysis and allow for many more options in terms of training and timing requirements [3]. Rapid sensory methods do not require trained panelists, as well as being less expensive, require less quantity of the samples of interest and are able to create a general product description [4]. Furthermore, rapid profiling methods are able to develop a vocabulary for the products of interest by asking consumers (also referred to as untrained partici-

María Ángeles Pozo-Bayón and Carolina Muñoz González (eds.), *Wine Analysis and Testing Techniques*, Methods and Protocols in Food Science, https://doi.org/10.1007/978-1-0716-3650-3_12,

pants) to evaluate the product [5]. Napping and sorting are two rapid profiling methods that have been used to evaluate the sensory properties of wine [6, 7].

Pagès introduced napping to the field of sensory evaluation [8]. Napping and projective mapping are very similar and napping can be considered a variation of the projective mapping method [9, 10]. Napping uses the same evaluation techniques as projective mapping by asking the participants to place samples within a two-dimensional (2D) space [11]. The participants are required to place the samples in the 2D space based on the similarities and differences of the samples. Samples that are similar are placed close together and those that are different are placed far apart. It allows the participants to assess the wine samples in a straightforward manner and leads to a graphical representation of the samples. Napping is also used paired with ultra-flash profiling (UFP), which asks the participants to provide descriptors of each sample [12]. The napping method is unique from projective mapping as the 2D space has to be rectangular (usually a 60 cm × 40 cm paper sheet which is approximately A2 size), data is not scaled, and the statistical analysis should be completed using Multiple Factor Analysis (MFA) [9]. Also, there are variations of napping called global and partial napping. The global method is a nonrestricted method and asks the participants to separate the samples based on the overall sensory modality. Global napping is considered the conventional version and allows the panelists to compare the samples on any modality they want [13]. In partial napping, the panelists are asked to concentrate on one sensory modality (appearance, aroma, taste, or texture) [14]. Furthermore, a variation of the Napping method has been combined with the sorting method [15]. Sorted napping replaces the UFP task with a sorting task.

The sorting task was introduced to the sensory evaluation field by Lawless et al. [16]. In the sorting task, participants are asked to group the samples based on their similarities [17]. The sorting task is usually referred to as the free sorting task and it asks the participants to evaluate the global perception of the products. Different types of panelists (trained and untrained participants) can be used during the sorting task, and it has been recommended that 20 panelists (trained or untrained) be included to provide interpretable results [18]. Also, the optimum number of samples for a sorting task is between 9 and 20 [19]. After the samples have been sorted, a description step (similar to UFP in the napping method) can take place to describe each group of the product and this referred to as labeled sorting [20]. The results of a sorting task are most commonly analyzed using multidimensional scale (MDS). MDS aims to create a map that indicates the similarities and differences between the different samples [2]. A sorting task is useful during the development of new wines (especially if there is a large number of different formulations) and is able to identify the relevant sensory

Table 1
Key components to be addressed before beginning sensory trial

Characteristics	Options
What type of participants should be included [21]?	Consumers Trained Experienced Expert
What type of sensory modality should be investigated [4]?	Global (overall) Partial (e.g., appearance, aroma, taste, mouthfeel)
How many samples should be included?	Optimum number in a napping task is 10–20 samples [22] Optimum number in a sorting task is 9–20 samples [19]
Use of a replicate sample to assess participant consistency [22]?	Yes No

properties of the wines that are important to consumers. Before starting a sensory trial using either method, some key questions need to be answered as outlined in Table 1.

This chapter will describe how to organize a napping and labeled sorting task to investigate the sensory properties of 10 red wines using consumers (untrained panelists).

2 Materials

In both the napping and sorting tasks described below, 12 wine samples were included (10 different wines and 2 blind replicates to check for individual consistency [23], **Note 1**). Approximately 25 ml of all the wines were served in pear-shaped transparent ISO glasses with coverslips. All glasses were labeled with three-digit random codes. All wine samples were presented at the same time and at the same temperature. Water and unsalted crackers were also presented as palate cleansers. All studies took place in individual sensory tasting booths under white light (**Note 2**).

3 Methods

3.1 Napping

1. Recruit wine consumers (untrained panelists- **Note 3**). Before beginning the trial be sure to receive informed consent from all participants.

2. Researchers should instruct the participants on how to conduct the napping task and the researchers may want to demonstrate the procedure for the participants by using other food items (**Note 4**).
3. The wine samples are then simultaneously presented (in a randomized order) to the participants and then participants were asked to complete the task on a 60 cm × 40 cm paper sheet (approximately A2 size). Participants should be presented with the instructions, "Please evaluate the wines and place them on the provided space according to how similar and dissimilar they are (based on your perception). The more similar the wines are, the closer they should be placed to each other, the more dissimilar they are the further apart they should be placed. If two products are placed close together, the more similar they are and the further apart, the more different they are. There are not right or wrong answers. When you are finished placing the products, write down the product codes in the appropriate locations on the sheet of paper and also please write 2-5 words to describe each product (**Note 5** and **6**). Please avoid product comparisons (product A is whiter than product B). Instead use terms such as low, medium, or high to express intensity." Participants should be encouraged to take as many sniffs or sips necessary to evaluate the wine. Also, between evaluating the samples participants should be instructed to use the provided palate cleansers (**Note 7**). Example of a participant's product map is shown in Fig. 1.
4. After the participants have completed the napping task, the product positions need to be measured from the bottom left corner (0.0 point) of the 2D space to the center of the written

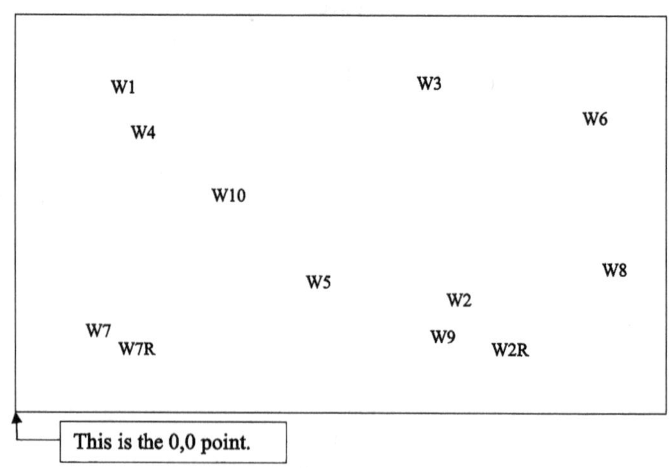

Fig. 1 An example of the 2D space after the participants have evaluated the 12 wine samples (ten samples and two replicates). The replicates are indicated by the R in the same name (e.g., W7R)

Sample	X1	Y1	X2	Y2	...	X50	Y51	Descriptor 1	Descriptor 2	...	Descriptor n
W1											
W2											
...											
W10											

Fig. 2 An example of how to organize the table based on the results of the napping task before using Multiple Factor Analysis. X and Y refer to the coordinates for the location of sample on the 2D space by each panelist [9]

code (Fig. 1). The zero point is the lower left corner of the 2D space (Fig. 1), and a table is created so that one row corresponds to the one sample (as seen in Fig. 2). Following the coordinates, the next columns correspond to descriptors from the UFP. The attributes are collected and summarized in a frequency table (Fig. 2) (**Note 8**).

5. Multiple Factor Analysis (MFA) is then used to produce bi-plots (Fig. 3) (**Note 9**). MFA normalizes each set of variables followed by principal component analysis (PCA) [24]. The PCA functions to analyze the variance from the different evaluations made by the participants and this is achieved by the using the measured x and y coordinates. The coordinates from each sample and each participant are rotated until a common trend is found [24]. These factors scores are used to create a plot that is representative of the similarities between participants' observations. MFA also calculates the frequency of descriptor citation for each product (**Note 10**). An example of the results of an MFA analyzing the results of a napping task on red wine is shown in Fig. 3.

3.2 Labelled Sorting

1. Like the napping task, wine consumers need to be recruited (untrained panelists- **Note 2**) and informed consent needs to be obtained.
2. The wines are then presented simultaneously to the participants (**Notes 11** and **12**). The wines should be arranged so that the presentation is ordered, and the first-order carry-over effects are balanced (**Note 13**).
3. Participants then look, smell, and taste the wines (based on the objective of the sensory trial—**Note 14**) and then sort the wines into as many groups as perceived by the participant, from a two to maximum of nine. The participants received the following instructions: "There are twelve red wines placed in front of you. Please, place the wines into groups that you consider alike, depending on your own criteria. You must create at least two groups and no more than eleven groups. Then, please write words that characterize each wine or group. Please take as much time as you want" (Fig. 4) (**Note 7**).

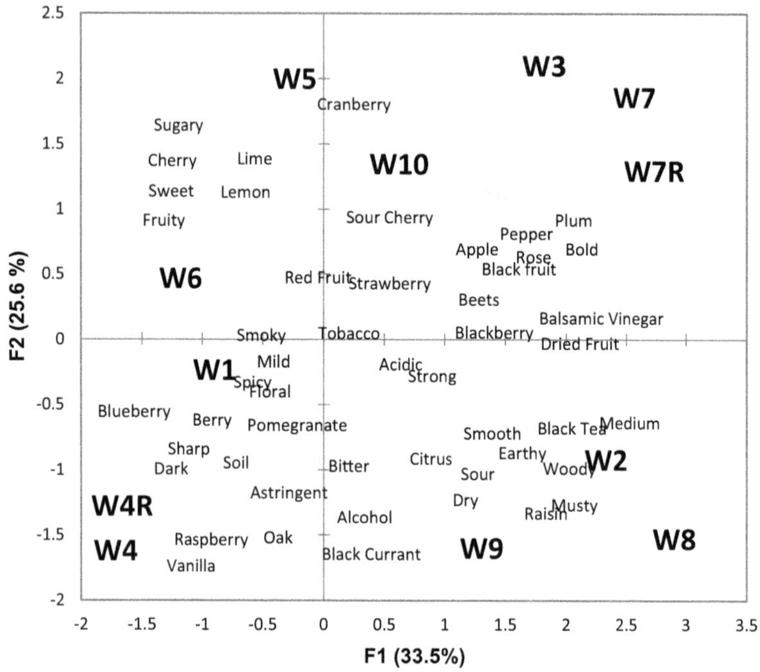

Fig. 3 An example of bi-plot produced from the first two dimensions of the Multiple Factor Analysis and the descriptors used to describe 12 wine samples (ten wines and two replicates)

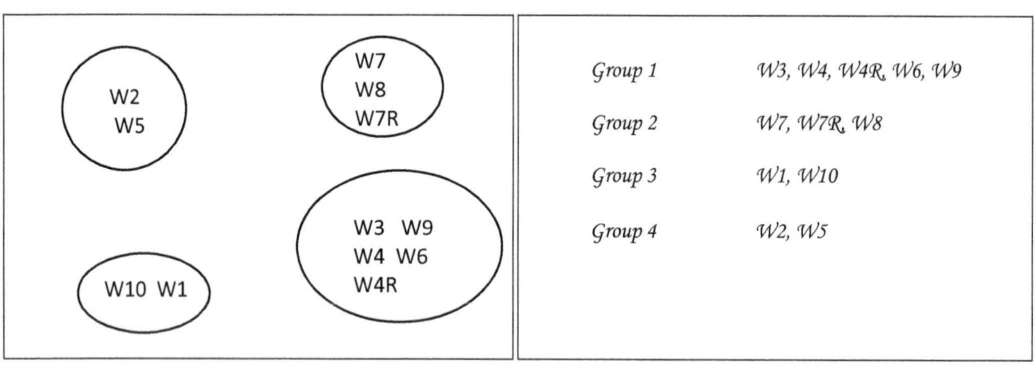

Fig. 4 Example of participants' scoring sheets

4. To analyze the results using MDS, the results from each participant are placed in an individual co-occurrence matrix. The value of 1 at the intersection of the row and column indicates that the participant grouped the wine samples together. A value of 0 indicates that the products were not placed together. Then the matrices from all of the different participants are then summed to obtain a global similarity matrix [25].

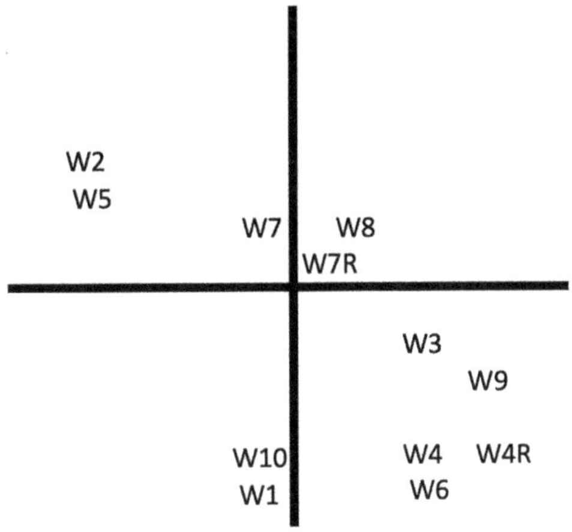

Fig. 5 Example of the 12 wine samples (10 wines and two replicates) on an MDS plot

5. Multidimensional scaling analyzes the similarity matrix. MDS is an analysis tool that can visualize proximities or distances between objects. Each point on the map (or plot) represents a wine sample and the points are arranged so that the wine samples that are alike to the participants are close together and those that are different are far apart (Fig. 5) (**Notes 15** and **16**) [25].
6. To analyze the descriptors associated with the groups of wine samples, a contingency table should be constructed. The products are placed in columns and the descriptors in rows. The frequency at which descriptors was used by the participants for each sample is placed in the contingency table. Generally, the descriptors with similar meanings are grouped and descriptors used less than a certain percentage (e.g., 10%) of assessors are discarded. The results from descriptors can then be projected onto the similarity plots by determining the correlations between the occurrence of descriptors and wine samples' factor scores (**Note 17**) [26].

4 Notes

1. The replicate sample or samples should be placed closely together on the 2D space by the participant.
2. If the partial napping or a sorting task is used to evaluate a specific sensory modality (e.g., aroma, taste, mouthfeel) except for evaluation of the appearance, then red lighting can be used.
3. Need to define what is meant by consumer. Screening criteria could include they do not have knowledge of sensory evaluation, wine, or work in a sensitive industry, had consumed wine

in the last 2 week and self-identified themselves to be regular wine drinkers. In addition, 50 consumers or more should be involved in a sensory trial using napping and 20 or more consumers for a labeled sorting task.

4. Beyond demonstrating the napping task, researchers have used a training exercise to demonstrate the method. For instance, Hopfer and Heymann [22] used paper shapes differing in color and asked the participants to position the figures on a provided space. The method could also be introduced using a video or by a brochure [27] for the participants.
5. Participants can also be presented with a list of sensory attributes to describe the wine (based on past evaluations or a literature review) to help them describe the sample. If a list is provided, it should be reinforced to the participants that can also use their own terms to describe the wines.
6. Napping task can be aimed at the global perception of the wine, but also could be aimed at a specific sensory modality (appearance, aroma, taste, mouthfeel, aftertaste) called partial napping. If using partial napping, then the researchers need to make it clear which sensory modality is of interest.
7. After completing the napping or sorting task, researchers may want to ask questions about wine knowledge, wine consumption habits, and demographics.
8. If a participant expresses an intensity of an attribute, each intensity is considered a separate attribute (e.g., sour, low sour, extremely sour are separate attributes).
9. Before conducting the MFA, some researchers remove descriptors if they do not meet a specific threshold. For instance, Hopfer and Heymman [22] removed the descriptor before the data analysis if it was not provided at least nine times by the participants.
10. MFA is the main statistical analysis tool to evaluate results from a napping task, but Hierarchical Multiple Factor Analysis and Generalized Procrustes Analysis have been used. Also, RV coefficients have been used to analyze the reproducibility of the results either from different participants or the same participants but different sessions of testing. Participant consistency can also be evaluated by using a people performance index [22].
11. You can include replicate samples to assess the reliability of the participant (i.e., Did they sort the replicate wine samples together?).
12. An issue with the sorting task (and napping task) is that all of the samples are presented at the same time, and it is difficult to control the temperature of the presented samples.

13. Participants can be instructed to follow the presentation order to ensure that each wine is tasted first an equal number of times.
14. Most sorting tasks involve the participants evaluating the global perception of the samples. But participants can be asked to focus on a certain sensory modality. Also, a Multiple Sorting task could be used where the participants perform the sorting task multiple times, and each time, they group the samples according to different criteria. The task is complete, when the participants cannot think of any criteria on which to sort the samples.
15. MDS can use different algorithms to visualize the results and the two main algorithms are metric and non-metric. A good overview of MDS and the different algorithms was written by Abdi [28].
16. Studies have also used Multiple Factor Analysis [29], Multiple Correspondence Analysis [30] and DISTATIS [30] to evaluate the results of a sorting task.
17. Contingency table can also be evaluated by correspondence analysis [30] or a multiple correspondence analysis [31].

References

1. Lestringant P, Delarue J, Heymann H (2019) 2010–2015: how have conventional descriptive analysis methods really been used? A systematic review of publications. Food Qual Prefer 71:1–7. https://doi.org/10.1016/j.foodqual.2018.05.011
2. Varela P, Ares G (2012) Sensory profiling, the blurred line between sensory and consumer science. A review of novel methods for product characterization. Food Res Int 48:893–908. https://doi.org/10.1016/j.foodres.2012.06.037
3. Liu J, Bredie WLP, Sherman E et al (2018) Comparison of rapid descriptive sensory methodologies: free-choice profiling, flash profile and modified flash profile. Food Res Int 106:892–900. https://doi.org/10.1016/j.foodres.2018.01.062
4. Louw L, Malherbe S, Naes T et al (2013) Validation of two napping® techniques as rapid sensory screening tools for high alcohol products. Food Qual Prefer 30:192–201. https://doi.org/10.1016/j.foodqual.2013.05.016
5. Sáenz-Navajas M-P, Alegre Y, de-la-Fuente A et al (2016) Rapid sensory-directed methodology for the selection of high-quality aroma wines. J Sci Food Agric 96:4250–4262. https://doi.org/10.1002/jsfa.7636
6. Perrin L, Symoneaux R, Maître I et al (2008) Comparison of three sensory methods for use with the napping® procedure: case of ten wines from Loire valley. Food Qual Prefer 19:1–11. https://doi.org/10.1016/j.foodqual.2007.06.005
7. Alegre Y, Sáenz-Navajas M-P, Ferreira V et al (2017) Rapid strategies for the determination of sensory and chemical differences between a wealth of similar wines. Eur Food Res Technol 243:1295–1309. https://doi.org/10.1007/s00217-017-2857-7
8. Pagès J (2005) Collection and analysis of perceived product inter-distances using multiple factor analysis: application to the study of 10 white wines from the Loire Valley. Food Qual Prefer 16:642–649. https://doi.org/10.1016/j.foodqual.2005.01.006
9. Dehlholm C, Brockhoff PB, Meinert L et al (2012) Rapid descriptive sensory methods – comparison of free multiple sorting, partial napping, napping, flash profiling and conventional profiling. Food Qual Prefer 26:267–277. https://doi.org/10.1016/j.foodqual.2012.02.012
10. Moss R, McSweeney MB (2022) Projective mapping as a versatile sensory profiling tool: a review of recent studies on different food

products. J Sens Stud 37:e12743. https://doi.org/10.1111/joss.12743

11. Risvik E, McEwan JA, Colwill JS et al (1994) Projective mapping: a tool for sensory analysis and consumer research. Food Qual Prefer 5: 263–269. https://doi.org/10.1016/0950-3293(94)90051-5
12. Liu J, Grønbeck MS, Di Monaco R et al (2016) Performance of flash profile and napping with and without training for describing small sensory differences in a model wine. Food Qual Prefer 48:41–49. https://doi.org/10.1016/j.foodqual.2015.08.008
13. Marcano J, Ares G, Fiszman S (2015) Comparison of partial and global projective mapping with consumers: a case study with satiating cheese pies. Food Res Int 67:323–330. https://doi.org/10.1016/j.foodres.2014.11.052
14. Kemp B, Pickering G, Willwerth J, Inglis D (2018) Investigating the use of partial napping with ultra-flash profiling to identify flavour differences in replicated, experimental wines. J Wine Res 29:302–309. https://doi.org/10.1080/09571264.2018.1532879
15. Pagès J, Cadoret M, Lê S (2010) The sorted napping: a new holistic approach in sensory evaluation. J Sens Stud 25:637–658. https://doi.org/10.1111/j.1745-459X.2010.00292.x
16. Lawless HT, Sheng N, Knoops SSCP (1995) Multidimensional scaling of sorting data applied to cheese perception. Food Qual Prefer 6:91–98. https://doi.org/10.1016/0950-3293(95)98553-U
17. Fleming EE, Ziegler GR, Hayes JE (2015) Check-all-that-apply (CATA), sorting, and polarized sensory positioning (PSP) with astringent stimuli. Food Qual Prefer 45:41–49. https://doi.org/10.1016/j.foodqual.2015.05.004
18. Chollet S, Lelièvre M, Abdi H, Valentin D (2011) Sort and beer: everything you wanted to know about the sorting task but did not dare to ask. Food Qual Prefer 22:507–520. https://doi.org/10.1016/j.foodqual.2011.02.004
19. Courcoux P, Qannari EM, Faye P (2015) Free sorting as a sensory profiling technique for product development. In: Delarue J, Lawlor JB, Rogeaux M (eds) Rapid sensory profiling techniques. Woodhead Publishing, pp 153–185
20. Bécue-Bertaut M, Lê S (2011) Analysis of multilingual labeled sorting tasks: application to a cross-cultural study in wine industry. J Sens Stud 26:299–310. https://doi.org/10.1111/j.1745-459X.2011.00345.x
21. Visalli M, Schlich P, Mahieu B et al (2023) First steps towards FAIRization of product-focused sensory data. Food Qual Prefer 104:104765. https://doi.org/10.1016/j.foodqual.2022.104765
22. Hopfer H, Heymann H (2013) A summary of projective mapping observations – the effect of replicates and shape, and individual performance measurements. Food Qual Prefer 28: 164–181. https://doi.org/10.1016/j.foodqual.2012.08.017
23. Nestrud MA, Lawless HT (2010) Perceptual mapping of apples and cheeses using projective mapping and sorting. J Sens Stud 25:390–405. https://doi.org/10.1111/j.1745-459X.2009.00266.x
24. Abdi H, Williams LJ, Valentin D (2013) Multiple factor analysis: principal component analysis for multitable and multiblock data sets. WIREs Comp Stats 5:149–179. https://doi.org/10.1002/wics.1246
25. Chollet S, Valentin D, Abdi H (2014) Free sorting task. In: Novel techniques in sensory characterization and consumer profiling, p 207–228
26. Abdi H, Valentin D (2007) Some new and easy ways to describe, compare, and evaluate products and assessors. In: New trends in sensory evaluation of food and non-food products. Vietnam National University – Ho Chi Minh City Publishing House, Ho Chi Minh, p 5–18
27. Veinand B, Godefroy C, Adam C, Delarue J (2011) Highlight of important product characteristics for consumers. Comparison of three sensory descriptive methods performed by consumers. Food Qual Prefer 22:474–485. https://doi.org/10.1016/j.foodqual.2011.02.011
28. Abdi H (2007) Metric Multidimensional Scaling. In: Salkind NJ (ed) Encyclopedia of measurement and statistics. Sage, Thousand Oaks, pp 598–605
29. Rodrigues JF, Mangia BA, e Silva JG et al (2020) Sorting task as a tool to elucidate the sensory patterns of artisanal cheeses. J Sens Stud 35:e12562. https://doi.org/10.1111/joss.12562
30. Cadoret M, Lê S, Pagès J (2009) A factorial approach for sorting task data (FAST). Food Qual Prefer 20:410–417. https://doi.org/10.1016/j.foodqual.2009.02.010
31. Valentin D, Chollet S, Lelièvre M, Abdi H (2012) Quick and dirty but still pretty good: a review of new descriptive methods in food science. Int J Food Sci Technol 47:1563–1578. https://doi.org/10.1111/j.1365-2621.2012.03022.x

Chapter 13

Rapid and Cost-Effective Methods for Wine Profiling: CATA/RATA

Dominique Valentin, Jeanne Brand, and Maria Pilar Sáenz-Navajas

Abstract

The objective of this chapter is to underline the main practical and methodological issues encountered when applying CATA or RATA to wine profiling. In the material section, we discuss some important methodological points that might impact the efficiency of the methodologies such as the type of assessors able to perform the task, the selection and presentation of the wines to be described, and the selection and presentation of the terms used to describe the wines. In the method section, we successively detail the CATA or RATA tasting procedures: session organization, instructions to be given to the assessors as well as statistical analysis.

Key words CATA, RATA, Rapid sensory methods, Wine profiling, Cost effective, Sensory analysis

1 Introduction

The Check-All-That-Apply (CATA) method and its Rate-All-That-Apply (RATA) variant are rapid profiling methods that can be used to assess sensory differences and similarities between wines and to highlight the main attributes of individual wines from the point of view of the wine tasters. These methods are based on a multiple-choice approach where assessors select the appropriate items from a list to best answer the question asked. First used in marketing research to understand consumer's opinions or motivations, CATA and RATA have been increasingly used in the last two decades for product sensory characterization. When conducted, a list of sensory attributes is compiled and sensory assessors should check the attributes, which best describe a set of products. Scores are then computed by counting the number of times a descriptor is selected for each product in the set to identify the most relevant attributes of each product. The main advantage of CATA is that it does not require deep cognitive processing, making it an easy profiling method both from the assessor's and experimenter's

María Ángeles Pozo-Bayón and Carolina Muñoz González (eds.), *Wine Analysis and Testing Techniques*, Methods and Protocols in Food Science, https://doi.org/10.1007/978-1-0716-3650-3_13,

points of view. Its main limitation is that frequencies are interpreted as an indicator of attribute intensities although they do not actually measure the intensity of perceived sensory attributes. To overcome this limitation an intensity-based variant of CATA named Rate-All-That-Apply (RATA) was proposed [1]. In RATA, assessors have to rate the intensity of the attributes they have checked in the CATA list. RATA is more cognitively demanding than CATA since assessors have to quantify the intensity of the attributes they perceive in the product, but, in counterpart, it has the advantage of limiting satisficing response strategies (checking attributes without thinking).

In the wine domain, the multiple-choice approach was first used under the name of citation frequency (CF) profiles by McCloskey et al. [2] and Le Fur et al. [3] to characterize the aromas of Chardonnay wines from California and Burgundy vineyards respectively. Later, Campo et al. [4] and Nanou et al. [5] used the citation frequency method to investigate the aroma properties of Spanish and Greek monovarietal white wines. The citation frequency profile is well adapted to describe wine aromas but it requires a training phase so that assessors are well acquainted with the attributes in the list. In the CATA method, the training phase is generally omitted, as the list of attributes is shorter than in the CF method and not limited to wine aromas. It generally includes simple attributes such as visual, olfactory, taste, or mouthfeel attributes that can be understood by untrained assessors. As a consequence, its popularity as a wine profiling tool increased recently. For example, Alencar et al. [6] used it to profile Syrah wine aged with oak chips, Vidal et al. [7, 8] to investigate the astringency of Tannat wines and Coste et al. [9] to characterize warm and cool climate dry red wine styles. The RATA variant has also been successfully applied to profile a large variety of wines as recently demonstrated by Rabitti et al. [10] with Italian wines.

2 Materials

2.1 Assessors

The number of assessors carrying out the sensory task depends on the method (CF, CATA, or RATA) and on the level of expertise of the participants.

2.1.1 Type of Assessors: Level of Expertise

The sensory task has to be carried out by a homogeneous group of assessors in terms of their experience with sensory analysis tasks and the product, wine in this case. Assessors can be classified as consumers, wine experts, and trained panelists attending to their level of expertise. Consumers are assessors that are not professionally involved in the wine industry and consume the product with a certain frequency (e.g., at least once a month). Wine experts are professionally involved in the wine industry and are defined as a person with extensive knowledge in wine acquired through deliberate practice and wine tasting (*see* **Note 1**). Trained panelists are assessors who attended a series of formal training sessions including the selection of the terms, their definitions, and development and presentation of standardized references or definitions for each term.

The choice of the type of assessor depends both on the methodology used (*see* **Note 2**) and the objective of the test. As a rule of thumb, CATA is more adapted for less experienced consumers as it is less cognitively demanding than RATA, which can present difficulties because of the use of scales. CF is generally used with trained assessors (*see* **Note 3**). Alternatively, experts can be selected to carry out CATA [8] or RATA [11] to both understand preference and get product characterization. The rationale behind this option is that experts are involved in the decision-making process of consumers at purchase [12], and they are able to be more discriminant than trained panels when using technical terms. They also present superior performance when describing wines, and thus they are expected to increase the discrimination among samples in comparison to consumers for complex products [13].

2.1.2 Number of Assessors

The number of assessors depends on the specific methodology as well as the objective of the study.

CF profile has been basically applied to characterize the aroma of wines with trained panelists using a large list of descriptors. Around 30 or 40 assessors are generally used with 30 being a minimum for the methodology to be reliable [14–16].

CATA and RATA were originally developed and applied to understand consumer perception [17–20], which requires a high number of assessors (min 60), especially when the degree of differences among products is small. The focus of these methodologies was then extended to the characterization of wine sensory properties. In this context, a short familiarization (semi-trained panel) or a full training (trained panel) phase can be used to reduce the number of assessors by increasing the consistency among them. For example, nine fully trained assessors were shown to be enough to successfully describe red wines [7] and 30 semi-trained assessors (2 h training) by CATA [21]. Likewise, with a short training (4–6 h training) the number of assessors can be reduced to 12–15 when using RATA to describe wines [10, 22] (*see* **Note 4**).

2.2 Wines

2.2.1 Number of Wines

Due to sensory fatigue, that could negatively influence a taster's perception and ability to identify wine sensory attributes, the number of wines presented in a tasting session should be considered. As a rule of thumb, CATA and CF profile allow the evaluation of a higher number of samples per session than RATA, as they are less cognitively demanding. Besides, wine experts are able to characterize a higher number of samples per session than consumers or trained panelists because they are used to evaluating a high number of wines in their everyday professional tasks. If wine industry professionals, such as winemakers and technical tasters can assess up to 22 wine samples in one session it is unwise to present more than 12 wine samples to consumers during a single tasting session. Yet the number of wines to be presented in a given session depends on the type of samples and the sensory category to be evaluated (i.e., color, aroma, flavor, taste, or mouthfeel).

Regarding the modalities to be evaluated, visual stimuli followed by orthonasal aroma description are less fatiguing than descriptions requiring to introduce wine in the mouth (i.e., flavor, taste, or mouthfeel). Thus, a priori a high number of wines can be presented for visual description, which will be limited only by the time availability of assessors. For orthonasal aroma characterization, the number of wines to be evaluated depends on both the type of assessors and the type of wines. Globally, if the wines are not too tannic, assessors seem to be able to evaluate overall in-mouth perception (i.e., aroma, taste, and mouthfeel) from 8 to 12 wines in a session with a break in the middle of the session to limit sensory fatigue. For example, trained and semi-trained assessors have been reported to be able to evaluate 9–12 wines per session by CF profile [4, 5] and 8 by RATA, respectively [22]. A 10-min break was enforced in the middle of each session (*see* **Note 5**). Consumers were shown to be able to evaluate the aroma, flavor, taste, and mouthfeel of 9 [23] or 10 red wines [24] per session using RATA (60 s between each wine sample was enforced) while trained panelists have shown discriminability and repeatability when describing these 3 modalities for up to 11 red wines per session applying an imposed pause of 10 min after 6 wines [25]. The number of wines to be evaluated in a given session decreases when the polyphenolic content of the wines increases. For example, for a CATA description focused on mouthfeel attributes with red wines with a high polyphenolic content such as Tannat wines, the number of wines was limited to 4 and 6 per session with trained panelists [7] and experts [8], respectively.

2.2.2 Sensory Diversity Among the Sample Set

CF profile, CATA, and RATA can be used indifferently to discriminate among wines presenting marked differences regardless of the level of expertise of assessors. This is less clear for very similar wines. RATA has been suggested to be able to discriminate among samples showing subtle differences (i.e., sharing attributes but differing in their intensity), and it seems to result in more stable wine and term configurations than CATA [26]. However, there is not a generalized agreement in the superiority of RATA over CATA in terms of discrimination ability among complex products, especially for descriptions carried out by consumers with wine sets showing low sensory diversity, as it is suggested to be dependent on the study and is sample specific [26].

2.2.3 Preparation of Wines

Prior to the sensory task, wine samples have to be screened by experimenters directly involved in the study (min 5 experienced assessors) in order to ensure that the sensory space of the sample set is suitable to assess the hypotheses of the study. In this bench session, experimenters have to confirm that samples that will be presented to the panel do not present cork taint (i.e., perceptible trichloroanisole-TCA: humidity-like aroma). Besides, it is probable that depending on the objective of the study, other odor and aroma-related defaults have to be screened and thus these samples avoided in the final description (*see* **Note 6**). Another point to consider when more than one bottle of wine is necessary for the sensory task is the bottle effect. This effect makes reference to differences presented by the same wine coming from different bottles. To avoid introducing this experimental effect, wine samples coming from different bottles can be mixed up and thus a homogeneous sample can be prepared. However, this practice has to be carried out with caution because a high amount of oxygen is dissolved in the wine when mixing bottles, which could lead to the evolution of wine samples if they have to be stored during the duration of the experiment (i.e., more than 2–3 days). In that case, the use of bottles freshly opened every day is recommended, and consequently, the bottle effect is assumed as it is expected to be lower than the evolution of wine flavor due to bottle mixing.

2.2.4 Presentation of Wines

The first point to consider when planning the session setting-up is the volume to be poured by glass. It usually ranges from 10 to 30 mL per glass. It is interesting to consider the volume to be poured based on the sensory modality to be evaluated. Accordingly, if samples are to be in-mouth characterized (i.e., flavor, taste, or mouthfeel), it is recommended to instruct the assessors to introduce the whole volume in the mouth to standardize the process inter- and intra-assessors. In that case, 10–15 mL is recommended. This restricted volume will also reduce the building-up effect (*see* **Note 7**) of in-mouth sensations, especially when dealing with high polyphenolic red wines. In any case, the use of rinsing agents is recommended between wine samples. While water and unsalted crackers are the most commonly used, pectin or yogurt can also be used as rinsing agents (*see* **Note 8**).

The second point to consider is the type of glass to be employed. The standardized ISO 3591:1977 glasses are recommended. They can be either clear or black glasses, these lasts are used to avoid the effect of visual cues on the perception of the rest of sensory modalities, which is especially relevant when the sample set presents a great variability of visual characteristics (i.e., turbidity or color) and the main aim of the task is to characterize the chemosensory properties of the sample set (*see* **Note 9**).

2.3 List of Terms

The selection of terms to be evaluated depends on the assessors, the method, and the sample set (*see* **Notes 1** and **2**).

In CF the list of attributes is generally long and assessors are asked to check only a limited number of attributes (around five in general). For example, in Le Fur, Mercurio, Moio, Blanquet, and Meunier [3], assessors were asked to check a maximum of six odor descriptors from a list of 78 terms, and in Campo, Do, Ferreira, and Valentin [4], they were asked to select between 2 and 5 terms in a list of 73 terms. Terms are generally organized in a hierarchical way inspired by the aroma wheel of Noble, Arnold, Buechsenstein, Leach, Schmidt, and Stern [27, 28] to simplify the visualization and processing of the terms and consequently reducing the cognitive charge. They are presented arranged either in a wheel or a table [4]. This structure usually comprises three levels or categories: family (most generic, e.g., fruit), subfamily (intermediate category, e.g., tropical fruit), or specific term (most specific, e.g., pineapple). CATA and RATA methods generally comprise a smaller number of terms than CF. For example, Vidal et al. [29] used a CATA list comprising 44 terms to assess how consumers describe astringency in red wine. An important point to consider when selecting the terms to be evaluated is the level of expertise of participants. While simple and easy-to-interpret terms are required for consumers, wine experts and trained panelists are able to use more complex terms. Yet, it has to be considered that the interpretation of the terms can strongly vary between experts and trained panels [8]. Table 1 provides some terms used in CF, RATA, or CATA attending to the expertise of the assessor.

3 Method

3.1 CATA

3.1.1 Session Organization

When CATA is performed, assessors receive the wine samples one at a time (monadically), blinded with 3-digit codes. The presentation order of the samples in a set is randomized across the assessors, in other words, each assessor will receive the samples in a different order according to a randomized design such as a Williams Latin-square design [31] (*see* **Note 10**).

3.1.2 Procedure

Assessors are asked to evaluate each wine and to check in the list the attributes that best describe the product. They can be asked to select as many options as they want (i.e., "check-all-that-apply") or they can be limited by being asked to only select the "*k*" most important or prominent options or attributes, this variation is referred to as "pick-*k*" attributes in the literature. When pick-*k* is used, assessors highlight the most important characteristics of the samples, where CATA will provide a more detailed comprehensive picture of the characteristics of the samples [31, 32]. Pick-*K* is well-suited for sensory profiling when experts (industry professionals)

Table 1
Terms used in CF, RATA, or CATA attending to the expertise of the assessor, the type of wine, and the type of wine

Assessor	Matrix	Method (Ref)	Terms
Consumers	White wine	CATA [18]	**Aroma/Flavor:** Fruity, tropical, floral, green, cooked vegetables, light, intense flavors **Taste:** Sweet, sour **Mouthfeel:** Tingly, sharp, dry, thick/viscous **Time-related:** Fades quickly **Others:** Typical sauvignon blanc
	Red wine	RATA [17]	**Appearance:** Brown, clarity, green, yellow **Aroma/Flavor:** Apple/pear, bubblegum, buttery, chemical, citrus, dried fruit, floral, green/grassy, herbaceous, honey, milky, nutty, petroleum, spice, stone fruits, sulfidic, toasted, tropical, wood, mineral, savory, **Taste:** Sour, bitter, salt, sweet **Mouthfeel:** Body, creamy, crisp, dry, spritz, watery **Time-related:** Length
Semi-trained panel	White, rosé, red wines	RATA [10]	**Aroma/Flavor:** Citrus, tropical fruits, red fruits, wild berries, fruit tree, nuts, dried/baked fruit, white floral, red floral, vegetative, balsamic, spices, earthy, roasted, woody, caramelized, yeast, ethereal, animal **Taste:** Sour, bitter, salty, sweet **Mouthfeel:** Body, astringency, alcohol
	White, red, oaked wines	RATA [22]	**Aroma/Flavor:** Citrus, tropical fruits, red/black fruits, white, yellow fruits, cooked/candied fruits, banana/amylic, floral, canned vegetables, cooked cabbage, rotten onion, black olive, oaked, smoky/roasted, lactic, camembert, rotten eggs, stagnant water/damp cloth, truffle, mineral, reduction, oxidation
Trained panel	Red wine	RATA [7, 30]	**Mouthfeel:** Dry, rough, aggressive, sand paper, pucker, harsh, abrasive, hard, coarse grain, irritant and complex, silky, fine emery, suede, mouthcoating, velvety, corduroy, adhesive, hard, soft, rich, full-body, green, grainy, satin **Time-related:** Persistent
	White, red wines	FC [5]	**Aroma/Flavor** (hierarchically structured): Citrus fruit (lemon, grapefruit, lime, orange), tropical fruit (melon, banana, pineapple), other fruit (apple, pear, peach, apricot), floral (citrus blossoms, rose, jasmine), vegetal/herbaceous (fresh-cut grass, mint, tea, chamomile), nuts (nuts), earthy (mushroom, earthy), caramelized (honey, caramel, vanilla), spicy (pepper)
Wine experts	Red wines	RATA[a] [11]	**Aroma/Flavor:** White fruit, yellow fruit, tropical fruit, red fruit, black fruit, dried fruit/compote, floral, alcohol, spices, roasted, vegetal, cooked vegetables, earthy, reduction, animal, citrus, nuts, balsamic, moldy, lactic **Taste:** Salty, bitter, sweet, sour **Mouthfeel:** Dry, dry on tongue side, dry on palate, sticky, dusty, grainy, sandy, coarse, unctuous, oily, fleshy, mouthcoating, silky, gummy, watery, burning, hot, prickly **Time-related:** Persistent

[a]The method used is a variant of RATA, Rate-K-attributes ($n = 5$). Participants could rate a maximum of 5 attributes

Table 2
Different randomized sample serving orders used for replicate tastings

Assessor	1st replicate			Break	2nd replicate		
	Position 1	Position 2	Position 3		Position 1	Position 2	Position 3
Assessor 1	Wine 1	Wine 2	Wine 3		Wine 2	Wine 1	Wine 3
Assessor 2	Wine 2	Wine 3	Wine 1		Wine 3	Wine 2	Wine 1
Assessor 3	Wine 3	Wine 1	Wine 2		Wine 1	Wine 3	Wine 2

perform the task, whereas CATA might be better suited for trained panelists who are used to describe wines in detail.

The terms on the CATA list are presented in different randomized orders for each assessor to avoid primacy effects (i.e., terms in the top part of the list are usually more frequently checked than the other terms) [33]. When terms corresponding to different sensory modalities are used it is recommended to organize the list of terms according to sensory modalities following the classical wine-tasting script (1) visual perception; (2) orthonasal perception (odors); (3) retronasal perception (aroma); (4) taste; (5) mouthfeel sensations, and (6) persistence [10].

When CATA is performed with consumers or technical experts and wine producers, each assessor only evaluates a wine once and no replicated tastings are performed. When a trained panel is used, repeatability and reproducibility is often assessed by performing the tasting sessions in replicate [4, 8, 14]. It is important to present the wines in a different randomized order when serving the second and/or third replicate to assessors, in other words, a specific assessor should not get the set of wines in the same order twice when tasting replicates. It is also wise to enforce a break between samples as well as between replicates when testing repeatability to minimize sensory fatigue (see Table 2).

3.1.3 Data Analysis

Overall panel performance can be assessed by calculating: (1) average reproducibility or repeatability indexes by using an average index (Ri) for a given "i" attribute as proposed by Campo et al. [4], a reproducibility index of the panelist "i" for sample "j" (RIij) or an overall reproducibility index of the panelist "i" (RIi) by averaging the RIij across samples [34], (2) attribute stability by the stability index (SIk), and (3) stability of the selection of terms by the SSIk index (*see* **Note 11**).

The data collected using CATA are coded in the form of individual matrices in which 1 corresponds to a checked term and 0 to an unchecked term (Fig. 1). A Cochran's Q test can be performed on these individual matrices to determine which attributes are significant [18, 35, 36]. Individual matrices can also be summed to obtain a frequency matrix (also called contingency

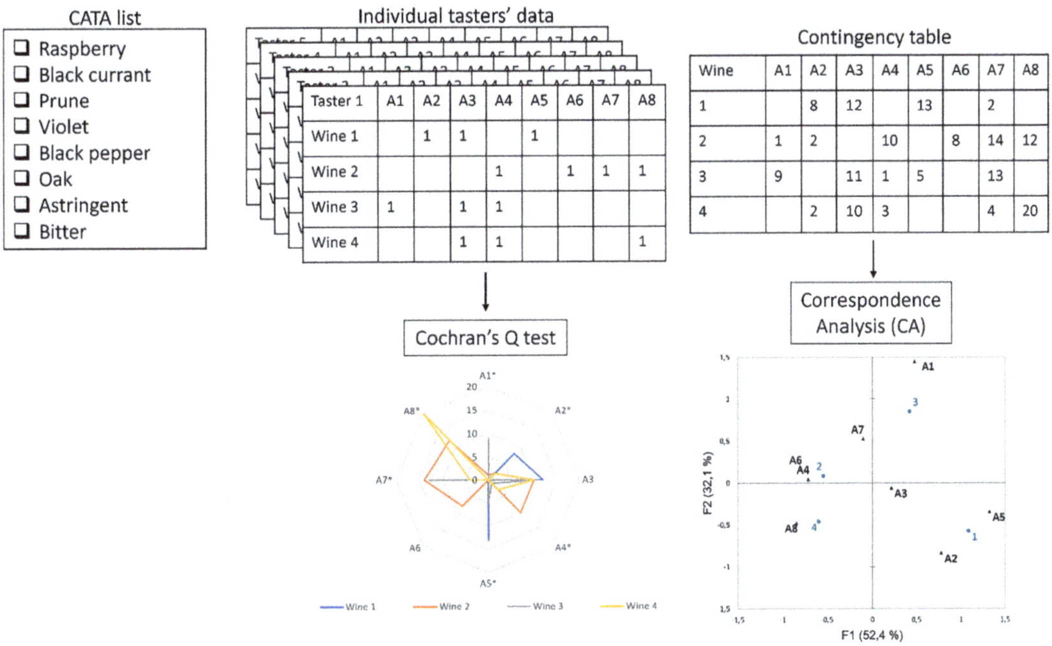

Fig. 1 Schematic representation of CATA data analysis

table) with the CATA list terms as the variables in the columns and the wines in the rows [14, 31]. This matrix is then submitted to a correspondence analysis (CA) to obtain a sensory map of the wines. CA is a generalized principal component analysis (PCA) suited for the analysis of frequency data [31].

The above-mentioned statistical analyses can be performed on all the CATA list terms or a set of reduced terms. The list of attributes could be reduced if terms with low citation frequencies are observed, in other words if few assessors used a term to describe the wines (*see* **Note 12**). The reduction of CATA list terms is used to address the fact that assessors might perceive the same "aroma" or "flavor" but select slightly different terms on the CATA list [4]. This could happen when the CATA list was set up to reflect subtle specific differences between samples, but the assessors could only perceive bigger differences or used slightly different vocabulary or concepts to describe similar perceptions. It was proposed in the literature to combine terms if less than 15% of the panel used a term [4, 14]. If a term with a low citation frequency could not be combined with another it is deleted prior to statistical analysis (*see* **Note 13**).

3.2 RATA

Same as for CATA.

3.2.1 Session Organization

3.2.2 Procedure

The main idea behind the development of RATA was to obtain intensity data rather than binary data, using simple scales that do not require training and aligning the assessors. As in CATA, assessors first indicate whether each term in the list applies to a given wine, and if so, they indicate the intensity of the descriptor. Various scales have been proposed and used with success ranging from 3-point to 9-point scales [17, 37–39] with a 15-point scale used as one of the first proposed [1]. The appropriateness of the scale used should be considered since a 3-point scale can, for example, be too limited to describe the intensity differences between similar wines in a set, whereas a 15-point scale can be too hard, due to the higher cognitive load that the assessor has to deal with (*see* **Note 14**).

3.2.3 Data Analysis

Various data analysis strategies have been proposed for RATA data. The first strategy is to analyze the citation frequencies of the terms as if the RATA data were CATA data, without taking the intensity rating data into account [33]. This strategy should be avoided since the discrimination ability of RATA decreases when RATA data are analyzed as binary data. A possible explanation is that a smaller difference can be better described using intensity rating than simple term selection or citation [16, 17]. The second strategy is to consider RATA data as ordinal data. The RATA scores (e.g., low, medium, and high) could be analyzed as weighted frequencies of citation. In this case, the numerical value 1 is assigned when "low" was marked, 2 for "medium," and 3 for "high." The sum of all the citation ratings for a term for a wine sample is computed. This is done for all the terms and for all the wines. Friedman's test can then be used to determine significant differences between the wines for each term and CA to visualize the wine space [26].

Alternatively, RATA data can be treated as continuous quantitative data as suggested by Meyners et al. [38]. In that case statistical analysis techniques such as analysis of variance (ANOVA) and PCA, routinely used for the analysis of descriptive sensory analyses like quantitative descriptive analysis (QDA), can be employed. To evaluate differences among wines, two-way ANOVAs with "assessor" as a random factor and "wine" as a fixed factor can be applied to RATA data [17, 22, 39]. If the "wine" factor is significant a post-hoc pairwise comparison test such as Bonferroni or Tukey's Honest Significant Difference can be used to investigate significant differences between the wines in terms of each RATA term [39]. If repetitions have been performed, three-way ANOVA with assessors, wines and replicates as fixed factors with first order

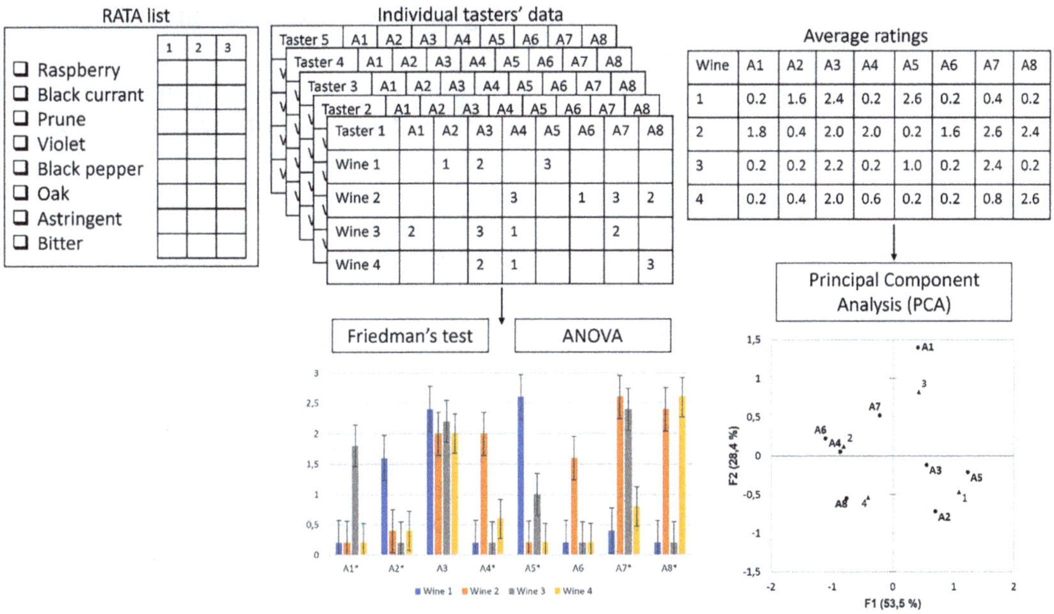

Fig. 2 Schematic representation of RATA data analysis

interactions can be calculated to evaluate panel performance. A sensory map of the wines can be obtained by applying a Principal Component Analysis (PCA) on the average ratings [22, 38, 39] (Fig. 2).

4 Notes

1. Different types of experts have been identified based on their developed skills through their training and practice, including winemakers, wine sellers (i.e., sommeliers and wine merchants), and wine critics [40]. It is important to consider that winemakers can provide more precise descriptors than sellers, while critics adopt an intermediate strategy by using both precise and global terms in wine descriptions [41].
2. The level of expertise of the assessors can result in different descriptions [8]. Consumers tend to be the least consensual, being able to discriminate among products based exclusively on simple easy-to-understand terms, while trained panels are the most discriminant as they share common concepts. Experts follow an intermediate approach, as they do not follow a standardized training but they attend wine tastings and formation courses regularly, which result in shared term conceptualization, especially among winemakers of the same region [42]. They can use complex terms consensually, being in certain cases even more discriminant than a trained panel,

especially when using complex, technical, and multidimensional terms [8].

3. One of the most difficult aspects of performing CATA and RATA is the generation and selection of the terms in the list [43]. Some pre-existing knowledge of the sensory characteristics of the samples to be evaluated, or at least an idea of the sensory space covered by the sample set, is needed to compile a list that will enable the researchers to obtain sufficient descriptions of the sensory characteristics of the samples and ensure that important attributes are not missed or left out, while performing the formal sensory evaluation. There are various strategies that can be employed in this regard. When little is known about the samples, existing lists or sensory wheels such as the aroma wheel of Noble et al. [27, 28] can be used. This often results in a long list of terms. In order to compile shorter lists that will be easier to use, descriptive analysis attributes, generated by a trained panel, during profiling of the samples to be analyzed or similar samples can be used. When descriptive analysis results are not available, focus group sessions conducted with consumers or technical experts, such as winemakers, can be scheduled to develop CATA or RATA lists product set specific lists.
4. The training stage can be also beneficial when the descriptors to be evaluated are complex. This phase will help to increase discrimination among samples. However, it has to be borne in mind that if the objective of the project is to understand consumer perception, training is not recommended [44].
5. Winemakers have been presented up to 22 red wines in one session to be described in terms of taste and mouthfeel [11] following rate-K-attributes, which is the homolog variant of CF for RATA. To this end, two flights of 11 wines were evaluated, with a compulsory break of 20 min between both flights and with a 10-min break every 5 samples).
6. However, the rejection of samples based on sensory defaults has to be carried out with caution, and it has to be considered that there is a great diversity of winemaking methods, and grape varieties, being the presence of defaults dependent on the style of the wine. For example, oxidation-related cues are characteristic and signals of quality in Sherry or Porto wines, while they are evident defaults in most common wines. Similarly, reduction aromas are sought in wines designed to carry out bottle aging, thus the matrix and the objective of the study has to be considered at this point of wine selection.
7. Perceived astringency and bitterness increase with repeated ingestions of wines, which contain astringent and bitter compounds, mainly polyphenols. This means that the last sample

will be perceived higher in astringency and bitterness than the first one. To avoid this build-up effect samples are presented in different order/position to each assessor. The use of rising agents also facilitates the dissipation of these accumulative sensations, as well as the imposition of pauses between wines.

8. The implementation of a water/pectin (1 g L^{-1}) protocol [45] between wines is recommended to avoid the building-up of certain sensations, especially of bitterness or astringent-related cues. Moreover, the use of natural yogurt has shown to be more effective than a solution of 2 g L^{-1} of pectin resulting in an increase of sample discrimination ability for astringency rating in Tannat wines [46]. In this context, it is important to evaluate the best rinsing agent depending on the study, as the ecology of the task can be committed because pectin or yogurt are meant to dissipate astringency, but they could ultimately influence the evaluation of other sensory modalities such as aroma or flavor.
9. An important consideration about the use of dark glasses is regarding its use when preference or quality is to be recorded by consumers or wine experts as reported by Parr et al. [47], and Valentin et al. [48]. The authors point out that there is a positivity effect of the clear glass condition over the dark glass. This has been firstly attributed to the fact that the dark glass seems to intimidate tasters, which leads them to be more cautious and conservative when scoring samples, and secondly, expectations and thus wine judgment could be modified as a result of cognitive and/or emotional positivity when employing clear glasses.
10. Randomization of sample presentation order across assessors using a Williams Latin Square design [49] is performed to minimize the carry-over effects from one sample to the next. An example of such a carry-over effect could be: if a very sweet sample is tasted first, the second sample might be scored lower in sweetness than when tasted after a sample that is not sweet at all. The bias caused by this phenomenon cannot be eliminated and is therefore minimized and balanced across the responses of all the assessors. A William Latin square design is balanced for first-order carry-over effects and is therefore appropriate to use. If an even number of samples is tested one Latin square is used and if the number of samples is uneven two Latin squares are used. When CATA/RATA is performed on six samples, and each sample is evaluated by every assessor, the following William Latin Square design can be used as serving order design (Table 3). Williams Latin Square designs can be generated by the "crossdes" R package (http://R-Project.org) written by Oliver Sailer sailer@statistik.uni-dortmund.de

Table 3
William Latin Square design for serving order design on six samples, considering that and each sample is evaluated by every assessor

Assessor	Position 1	Position 2	Position 3	Position 4	Position 5	Position 6
Assessor 1	Wine 1	Wine 2	Wine 6	Wine 3	Wine 5	Wine 4
Assessor 2	Wine 2	Wine 3	Wine 1	Wine 4	Wine 6	Wine 5
Assessor 3	Wine 3	Wine 4	Wine 2	Wine 5	Wine 1	Wine 6
Assessor 4	Wine 4	Wine 5	Wine 3	Wine 6	Wine 2	Wine 1
Assessor 5	Wine 5	Wine 6	Wine 4	Wine 1	Wine 3	Wine 2
Assessor 6	Wine 6	Wine 1	Wine 5	Wine 2	Wine 4	Wine 3

11. Control indexes for monitoring the performance of panels in CATA methodology are schematized in Table 4.
12. The term reduction strategy is typically used when a trained panel or wine experts perform CATA [4, 14]. These panels consist of fewer assessors than when consumer panels perform CATA. Hence, lower citation frequencies occur and often reduction of the terms is necessary to obtain meaningful results.
13. The reason for deleting terms with a low citation frequency is that in CA the weight assigned to each term is inversely proportional to its use. If these terms are kept in the analysis, they might contribute strongly to the wine space when in fact they are rather anecdotal.
14. Regarding the type of scale, there are publications employing 3-(low-medium-high) [37] *4*, 5- (1 = slightly applicable and 5 = very applicable) [38], 7- (1 = extremely low; 7 = extremely high) [17] or 9-point scales (1 = low; 9 = high) [39] or even 15-cm unstructured scales [1]. In overall, wine experts and semi-trained panelists have successfully used the 7-point scale [11, 22], and fully trained panelists do not have problems as long as they are trained in the use of the selected scale. However, there is not a generalized consensus on the best scale to be used by consumers. A limited range of the scale (i.e., 3 points) can limit the discriminability of samples; however, increasing the number of points should be considered with caution, because the higher number of points, the higher the cognitive load, which leads to infringe on the main principle of consumer-based approaches: simple and easy to use for assessors. Danner, Crump, Croker, Gambetta, Johnson, and Bastian [17] suggested that the 7-point scale is suitable for consumers as it reaches the compromise between the capacity of discrimination and the simplicity of its use.

Table 4
Control indexes for monitoring performance of panel in CATA methodology

Index	Codes	Interpretation
RIj: Average reproducibility index for an i panelist $Ri = \frac{1}{J} * \sum_{j=1}^{j=J} \left(2 * \frac{ter_{comj}}{ter_{rep1j} + ter_{rep2j}}\right)$	ter_{comj}: n° of common terms used by a panelist i in two identical wines 1 and 2 ter_{rep1j}/ter_{rep2j}: n° of terms employed to describe wines 1 and 2 J: n° of wines	Repeatability/reproducibility control for individual panelists Range: 0 (lack of repeatability/ reproducibility) to 1 (perfect repeatability/ reproducibility)
RIij reproducibility index for an i panelist for a wine j $RI_{ij} = \frac{ter_{idij}}{ter}$ RIi: Global reproducibility index for an i panelist $RIi = \text{average}(RI_{ij})$	ter_{idij}: n° of common terms used by an i panelist to describe two identical j wines *ter*: n° of total terms in the global list.	Repeatability/reproducibility control for individual panelists Range: 0 (lack of repeatability/ reproducibility) to 1 (perfect repeatability/ reproducibility)
SI_k: Stability index of a k attribute $SIk = \frac{100}{J*N} \sum_{n=1}^{n=N} (sam_{nk})$	sam_{nk}: n° of wines in which the panelist n used the term k similarly in two different sessions N: n° of panelists J: n° of wines	Stability control of the k term Percentage of panelists that employed the term similarly to describe the simple wine in different sessions Range: 0 (lack of stability) to 100 (perfect stability)
SSI_k: Stability index in the selection of a k attribute $SSI_k = \frac{200 * \sum_{n=1}^{n=N} ssam_{nk}}{\sum_{n=1}^{n=N} (s1_{nk}) + \sum_{n=1}^{n=N} (ss2_{nk})}$	sam_{nk}: n° of wines in which the panelist n used the term k identically in two sessions $S1_{nk}$ and $S2_{nk}$: n° of wines in which the panelist n used the attribute k in sessions 1 and 2, respectively	Stability control in the selection of the k term Percentage of panelists that used the term identically for describing the same simple in different sessions in relation to the average of panelists that used the term in at least one out of the two sessions
RIj: Average reproducibility index for an i panelist $Ri = \frac{1}{J} * \sum_{j=1}^{j=J} \left(2 * \frac{ter_{comj}}{ter_{rep1j} + ter_{rep2j}}\right)$	ter_{comj}: n° of common terms used by a panelist i in two identical wines 1 and 2 ter_{rep1j}/ter_{rep2j}: n° of terms employed to describe wines 1 and 2 J: n° of wines	Repeatability/reproducibility control for individual panelists Range: 0 (lack of repeatability/ reproducibility) to 1 (perfect repeatability/ reproducibility)
RIij reproducibility index for an i panelist for a wine j $RI_{ij} = \frac{ter_{idij}}{ter}$ RIi: Global reproducibility index for an i panelist $RIi = \text{average}(RI_{ij})$	ter_{idij}: n° of common terms used by an i panelist to describe two identical j wines *ter*: n° of total terms in the global list.	Repeatability/reproducibility control for individual panelists Range: 0 (lack of repeatability/ reproducibility) to 1 (perfect repeatability/ reproducibility)

References

1. Reinbach HC, Giacalone D, Ribeiro LM, Bredie WLP, Frøst MB (2014) Comparison of three sensory profiling methods based on consumer perception: CATA, CATA with intensity and napping®. Food Qual Pref 32:160–166
2. McCloskey LP, Sylvan M, Arrhenius SP (1996) Descriptive analysis for wine quality experts determining appellations by chardonnay wine aroma. J Sens Stud 11:49–67
3. Le Fur Y, Mercurio V, Moio L, Blanquet J, Meunier JM (2003) A new approach to examine the relationships between sensory and gas chromatography – Olfactometry data using generalized procrustes analysis applied to six French chardonnay wines. J Agric Food Chem 51:443–452
4. Campo E, Do BV, Ferreira V, Valentin D (2008) Aroma properties of young Spanish monovarietal white wines: a study using sorting task, list of terms and frequency of citation. Aus J Grape Wine Res 14:104–115
5. Nanou E, Mavridou E, Milienos FS, Papadopoulos G, Tempère S, Kotseridis Y (2020) Odor characterization of whitewines produced from indigenous Greek grape varieties using the frequency of attribute citation method with trained assessors. Foods 9:1396
6. Alencar NMM, Ribeiro TG, Barone B, Barros APA, Marques ATB, Behrens JH (2019) Sensory profile and check-all-that-apply (CATA) as tools for evaluating and characterizing Syrah wines aged with oak chips. Food Res Int 124: 156–164
7. Vidal L, Antúnez L, Giménez A, Medina K, Boido E, Ares G (2017) Sensory characterization of the astringency of commercial Uruguayan Tannat wines. Food Res Int 102:425–434
8. Vidal L, Antúnez L, Giménez A, Medina K, Boido E, Ares G (2018) Astringency evaluation of Tannat wines: comparison of assessments from trained assessors and experts. J Sens Stud 33:e12330
9. Coste A, Sousa P, Malfeito-Ferreira M (2018) Wine tasting based on emotional responses: an expedite approach to distinguish between warm and cool climate dry red wine styles. Food Res Int 106:11–21
10. Rabitti NS, Cattaneo C, Appiani M, Proserpio C, Laureati M (2022) Describing the sensory complexity of Italian wines: application of the Rate-All-That-Apply (RATA) method. Foods 11:2417
11. Sáenz-Navajas MP, Ferrero-del-Teso S, Jeffery DW, Ferreira V, Fernández-Zurbano P (2020) Effect of aroma perception on taste and mouthfeel dimensions of red wines: correlation of sensory and chemical measurements. Food Res Int 131:108945
12. Marin AB, Durham CA (2007) Effects of wine bottle closure type on consumer purchase intent and price expectation. Am J Eno Vitic 58:192–201
13. Hughson AL, Boakes RA (2002) The knowing nose: the role of knowledge in wine expertise. Food Qual Pref 13:463–472
14. Campo E, Ballester J, Langlois J, Dacremont C, Valentin D (2010) Comparison of conventional descriptive analysis and a citation frequency-based descriptive method for odor profiling: an application to Burgundy pinot noir wines. Food Qual Pref 21:44–55
15. Sáenz-Navajas MP, Gonzalez-Hernandez M, Campo E, Fernández-Zurbano P, Ferreira V (2012) Orthonasal aroma characteristics of Spanish red wines from different price categories and their relationship to expert quality judgements. Aus J Grape Wine Res 18:268–279
16. Sáenz-Navajas MP, Ballester J, Pêcher C, Peyron D, Valentin D (2013) Sensory drivers of intrinsic quality of red wines. Effect of culture and level of expertise. Food Res Int 54: 1506–1518
17. Danner L, Crump AM, Croker A, Gambetta JM, Johnson TE, Bastian SEP (2018) Comparison of rate-all-that-apply and descriptive analysis for the sensory profiling of wine. Am J Eno Vitic 69:12–21
18. Ares G, Antúnez L, Bruzzone F, Vidal L, Giménez A, Pineau B et al (2015) Comparison of sensory product profiles generated by trained assessors and consumers using CATA questions: four case studies with complex and/or similar samples. Food Qual Pref 45: 75–86
19. Souza Gonzaga L, Bastian SEP, Capone DL, Danner L, Jeffery DW (2022) Consumer perspectives of wine typicity and impact of region information on the sensory perception of cabernet sauvignon wines. Food Res Int 152: 110719
20. Nguyen ANH, Johnson TE, Jeffery DW, Capone DL, Danner L, Bastian SEP (2020) Sensory and chemical drivers of wine consumers' preference for a new shiraz wine product containing Ganoderma lucidum extract as a novel ingredient. Foods 9:224
21. Veríssimo CM, de Macêdo MS, de Andrade Lima LL, Pereira GE, Maciel MIS (2021) A short training as an enhancer of sensory ability:

the case of red wine consumers. J Sens Stud 36: e12629

22. Franco-Luesma E, Sáenz-Navajas MP, Valentin D, Ballester J, Rodrigues H, Ferreira V (2016) Study of the effect of H_2S, MeSH and DMS on the sensory profile of wine model solutions by Rate-All-That-Apply (RATA). Food Res Int 87:152–160
23. Copper AW, Johnson TE, Danner L, Bastian SEP, Collins C (2019) Preliminary sensory and chemical profiling of Cypriot wines made from indigenous grape varieties Xynisteri, Maratheftiko and Giannoudhi and acceptability to Australian consumers. Oeno One 53:229–248
24. Mezei LV, Johnson TE, Goodman S, Collins C, Bastian SEP (2021) Meeting the demands of climate change: Australian consumer acceptance and sensory profiling of red wines produced from non-traditional red grape varieties. Oeno One 55:29–46
25. Ferrero-del-Teso S, Arias I, Escudero A, Ferreira V, Fernández-Zurbano P, Sáenz-Navajas MP (2020) Effect of grape maturity on wine sensory and chemical features: the case of Moristel wines. LWT 118:108848
26. Ares G, Bruzzone F, Vidal L, Cadena RS, Giménez A, Pineau B et al (2014) Evaluation of a rating-based variant of check-all-that-apply questions: rate-all-that-apply (RATA). Food Qual Pref 36:87–95
27. Noble AC, Arnold RA, Masuda BM, Pecore SD, Schmidt JO, Stern PM (1984) Progress towards a standardized system of wine aroma terminology. Am J Eno Vitic 35:107–109
28. Noble AC, Arnold RA, Buechsenstein J, Leach EJ, Schmidt JO, Stern PM (1987) Modification of a standardized system of wine aroma terminology. Am J Eno Vitic 38:143–146
29. Vidal L, Giménez A, Medina K, Boido E, Ares G (2015) How do consumers describe wine astringency? Food Res Int 78:321–326
30. Rinaldi A, Vecchio R, Moio L (2021) Differences in astringency subqualities evaluated by consumers and trained assessors on Sangiovese wine using Check-All-That-Apply (CATA). Foods 10:218
31. Valentin D, Chollet S, Lelievre M, Abdi H (2012) Quick and dirty but still pretty good: a review of new descriptive methods in food science. Int J Food Sci Tech 47:1563–1578
32. Chollet S, Valentin D (2000) Le degré d'expertise a-t-il une influence sur la perception olfactive? Quelques éléments de réponse dans le domaine du vin. Annee Psychol 100:11–36
33. Ares G, Jaeger SR (2013) Check-all-that-apply questions: influence of attribute order on sensory product characterization. Food Qual Pref 28:141–153
34. Jaeger SR, Chheang SL, Yin J, Bava CM, Gimenez A, Vidal L et al (2013) Check-all-that-apply (CATA) responses elicited by consumers: within-assessor reproducibility and stability of sensory product characterizations. Food Qual Pref 30:56–67
35. Manoukian EB (1986) Mathematical nonparametric statistics. Gordon & Breach, New York
36. Meyners M, Castura JC (2014) Check-all-that-apply questions. In: Varela P, Ares G (eds) Novel techniques in sensory characterization and consumer profiling, 1st edn. CRC Press, Boca Ratón
37. Symoneaux R, Patron C, Neethling E, Coulon-Leroy C (2022) HRATA: a new sensory methodology using advantage of wine aromatic wheels. In: 5th International conference series on Wine Active compounds (WAC). IVES Conference Series, Dijon
38. Meyners M, Jaeger SR, Ares G (2016) On the analysis of Rate-All-That-Apply (RATA) data. Food Qual Pref 49:1–10
39. Oppermann AKL, de Graaf C, Scholten E, Stieger M, Piqueras-Fiszman B (2017) Comparison of Rate-All-That-Apply (RATA) and descriptive sensory analysis (DA) of model double emulsions with subtle perceptual differences. Food Qual Pref 56:55–68
40. Otheguy M, Honoré-Chedozeau C, Valentin D (2021) Do wine experts share the same mental representation? A drawing elicitation study with wine makers, sellers, and critics. Food Qual Pref 94:104302
41. Otheguy M (2022) La face cachée de l'expertise! Représentations et descriptions des vins par différents groupes d'experts au sein d'un vignoble en pleine évolution. Université de Bourgogne, le Beaujolais
42. Sáenz-Navajas MP, Arias I, Ferrero-del-Teso S, Escudero A, Ferreira V, Fernández-Zurbano P et al (2021) Access to wine experts' long-term memory to decipher an ill-defined sensory concept: the case of green wine. Oeno One 55:69–79
43. Lazo O, Claret A, Guerrero L (2016) A comparison of two methods for generating descriptive attributes with trained Assessors: Check-All-That-Apply (CATA) vs. Free Choice Profiling (FCP). J Sens Stud 31:163–176
44. Ares G, Picallo A, Coste B, Antúnez L, Vidal L, Giménez A et al (2018) A comparison of RATA questions with descriptive analysis: insights from three studies with complex/similar products. J Sens Stud 33:e12458

45. Colonna AE, Adams DO, Noble AC (2004) Comparison of procedures for reducing astringency carry-over effects in evaluation of red wines. Aus J Grape Wine Res 10:26–31
46. Vidal L, Antúnez L, Giménez A, Ares G (2016) Evaluation of palate cleansers for astringency evaluation of red wines. J Sens Stud 31:93–100
47. Parr WV, Grose C, Hedderley D, Medel Maraboli M, Masters O, Araujo LD et al (2020) Perception of quality and complexity in wine and their links to varietal typicality: an investigation involving pinot noir wine and professional tasters. Food Res Int 137:109423
48. Valentin D, Parr WV, Peyron D, Grose C, Ballester J (2016) Colour as a driver of pinot noir wine quality judgments: an investigation involving French and New Zealand wine professionals. Food Qual Pref 48:251–261
49. Williams EJ (1949) Experimental designs balanced for the estimation of residual effects of treatments. Australian J Sci Res A2:149–168

Chapter 14

Time-Intensity Methodology for Wine Flavor Evaluation

Celia Criado, Carolina Muñoz González, and María Ángeles Pozo-Bayón

Abstract

Traditionally, sensory methods such as quantitative descriptive analysis consider the sensory properties like a static phenomenon. However, due to the dynamic nature of wine consumption, sensory dynamic methodologies provide us with more specific information about the evolution of the flavor perception of a wine, also considering the long-lasting perception experienced when the wine is already consumed. In this sense, the Time-Intensity methodology evaluates how the intensity of an attribute varies over time. The purpose of this chapter is to provide the necessary steps to perform a Time-Intensity analysis for wine flavor evaluation, also showing how to train the panel and how data should be collected and processed.

Key words Time-Intensity, Dynamic sensory perception, Sensory analyses, Trained panel

1 Introduction

Sensory techniques like quantitative descriptive analysis (QDA) consider sensory properties as fixed attributes. Nonetheless, during wine consumption, the release of flavor compounds occurs gradually over time, following a non-constant rate. This is why flavor perception during wine tasting is a dynamic process, which evolves over time, also including the long-lasting perception (persistence) once the wine has been ingested [1].

To overcome the limitations of QDA, dynamic methodologies were developed in order to have a better understanding of the dynamic changes in flavor perception during the consumption of food and beverages. Currently, there are different dynamic sensory methodologies (Time-Intensity, Temporal Check-All-That-Apply, Temporal Dominance Sensations) [2–6]. Among them, Time-Intensity (TI) methodology measures the temporal changes in the sensory attributes of a product over time [7]. It involves recording and obtaining the variations in speed, persistence, and intensity of a single specific sensory attribute during the evaluation of the product [8]. The TI method offers an advantage by enabling a thorough evaluation with a consistent focus on a specific attribute. This

María Ángeles Pozo-Bayón and Carolina Muñoz González (eds.), *Wine Analysis and Testing Techniques*,
Methods and Protocols in Food Science, https://doi.org/10.1007/978-1-0716-3650-3_14,

method yields accurate outcomes in the analysis of sensations during food tasting, particularly in capturing the evolution of mouthfeel and flavor perception [9, 10].

The main disadvantage of this methodology is that it requires a certain degree of training of the panel to obtain accurate results; for instance, the panel need to be trained in the recognition of the quality and intensity of the attribute/s to be assessed and in the performance of the TI methodology. Recent tools, such as TI Reliability Index (TI-RI) allow us to explore the inter-individual variation of the panel, which is an important source for data variability [10].

TI is mainly directed for the evaluation of a single attribute per analysis. However, there are two variations of this method, in which two or more sensory attributes can be measured simultaneously. For instance, the Dual Attribute Time-Intensity (DATI) [3] or the Multiple Attribute Time-Intensity (MATI) methods, can be used to evaluate two or more attributes simultaneously. The DATI method consists in the simultaneous evaluation of the perception of two sensory attributes (e.g., sweetness and peppermint aroma) [3]. In contrast, the MATI methodology consists of allowing multiple attributes to be compiled intermittently using a rhythm through repeated cycling of the attributes with specific timed events captured over the course of a run [4]. The main benefit of these two variants is the simultaneous analysis of multiple attributes, which helps to reduce the number of evaluations required for each product. However, evaluating two attributes concurrently can lead to increased fatigue for the panelists. By utilizing DATI and MATI, the number of assessments per session can be reduced. Nevertheless, it is worth noting that simultaneous assessment can be challenging, necessitating additional training sessions. In addition, MATI does not collect all continuous scores for each attribute, reducing the accuracy of TI [11].

Since most dynamic sensory methodologies are based on TI, and because is still a widely used dynamic sensory approach [12], in the following sections, the step-by-step procedure for performing a TI evaluation of wine flavor attributes will be described.

2 Materials

2.1 Sensory Panel

Sensory panel should meet several criteria: number of volunteers, age (homogeneous age group and consider the minimum age of access to alcoholic beverages), non-smokers, free from diseases or allergies related to wine components and not pregnant women (*see* **Notes 1** and **2**).

All volunteers should be informed of the nature of the study and should complete a written consent form prior their participation in the study. Furthermore, the involvement of volunteers is crucial due to the complexity of the assessment process, and it is essential that they are enough motivated to participate [13].

As Time-Intensity analysis is a quantitative method, the use of scales requires trained panelists. TI requires great concentration as the panelist must focus on the changes in the flavor intensity perception of the specific attribute over a designated time span. Therefore, this technique requires additional training compared to others [14]. The number of panelists might vary depending on the study's objective, nonetheless, a trained panel should have a minimum of 10 panelists.

To facilitate the training, it is recommended to recruit volunteers who are familiar with the food to be analyzed, in this case, with wine. In fact, it is advisable that all the recruited panelists have a similar experience with wine in order to facilitate the homogeneous work of the panel. Nonetheless, several training sessions are advisable (*see* Subheading 3.2.).

2.2 Software

The use of specific software for data collection is very recommended. Most software allows to obtain the TI curves for a visual analysis together with the analysis of the curve parameters (*see* Subheading 3.3.1.). Some software that include sensory TI analysis are: FIZZ (Suffolk, England) [15], SIMS2000 (New Jersey, USA), Compusense Cloud (Compusense Inc., Guelph, Canada) [12] or Eye Question (Elst, Netherlands), SENSEBIT (Lugo, Spain) among others. TI software can be used via computers, tablets, and even mobile phones.

2.3 Wine Samples

For wine flavor analysis by TI methodology, samples should be served in ISO standard wine glasses, preferably opaque to avoid color bias. These wine glasses should be labeled with a random three-digit code. In addition, the order of presentation of the samples will be randomized among the panelists.

Another point to consider is the amount of wine consumed. This parameter must be the same in each test, so it must be measured. In general, the amount of wine should be enough to perform a zip of 10–15 mL (for swallowing or mouth rinsing).

The temperature of the wine should be always the same. If the wines are refrigerated, it is recommended to prepare them always at the same time in advance. Finally, the wine glasses should be covered with Petri dishes to avoid volatile lost.

A glass of water and/or palate cleansers such as unflavored crackers should be available in all sessions to clean the mouth between samples avoiding flavor-carry over effects.

3 Method

3.1 Wine Consumption Protocol

The dynamic TI methodology requires a complete protocol with instructions for the panelists (Fig. 1). This protocol has to include the procedures for wine consumption and attribute assessment using the selected software. Small variations in the performance of the test will cause high variation among panelists, thus decreasing the accuracy of the analysis. This protocol should be adapted according to the selected software. Training the panelists in the protocol is essential.

One of the main guidelines is the rinsing time. A soft rinse (usually of 30 s) is recommended in order to favor the equilibration of the flavor compounds within the oral cavity (*see* **Note 3**) [16]. After the wine rinses, panelists should be very focused on the task maintaining the lips closed and avoiding flavor losses by minimizing distractions throughout the assessment.

3.2 Training

To analyze an attribute using the TI methodology, it will be necessary to train the panel in different aspects (*see* **Note 4**). Initially, you will start with training them in the recognition of the flavor attributes. Flavor training can be carried out using the same wine or a synthetic wine matrix with added chemical odorant molecules at different concentrations. Some of the tests that can be used during training recognition are triangular (sample with the attribute versus sample without the attribute) or paired comparison tests.

During tasting, and once the wine has been swallowed or such in some tests, after mouth-rinsing and expectoration, the perceived intensity of the sensory attributes will change after the first exhalation. Therefore, once the panelists recognize the attribute, they should assess whether they are also able to differentiate the attribute at different concentrations. To do so, it is common to use three different concentrations of the flavor attribute followed by a concentration ranking test. During the ranking test, several samples (numerically coded) of *different concentrations of the flavor attribute* will be presented and the panelists will be asked to rank them in order of highest to lowest concentration.

Once checked the performance of the panel in the recognition of the sensory attributes at different concentrations, it is necessary a training in the TI methodology, to rate *the evolution of the flavor attribute over time* (*see* **Note 5**). This training ends when there is around 40% difference between the replicates of the sample considering the same individual [8, 17]. It is important to note, however, that greater the level of training the panel receives, the higher the level of consistency and reliability in the obtained results (*see* **Note 6**).

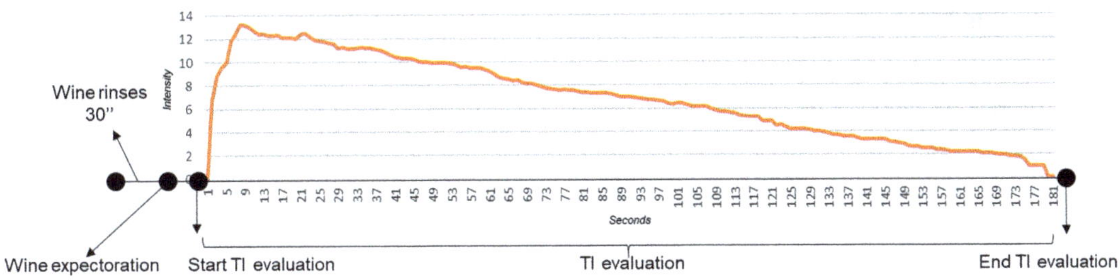

Fig. 1 Graphical example of the Time-Intensity assessment procedure

3.3 Time-Intensity Evaluation

The evaluation should be done in different sessions than the training sessions (*see* **Note 7**). The optimization and programming of the sessions could be different in the different software (*see* Subheading 3.3.1). However, there are some common aspects related to the TI evaluation. Like this, the evaluation of the attribute begins when the panelists press the Start button (in some cases it appears with the symbol "play" and in other cases a button with the word "start") (Fig. 2). From this moment the timer is activated and the program begins to record all the data that score on a scale from 0 (not present) to 100 (maximum intensity). The panelists must indicate all the changes that they perceive in the intensity of the sensory attribute during the setting time. The assessment ends when the timer reaches the setting time [18]. During the assessment, the panelists should be told if and when they are allowed to swallow (*see* **Note 8**).

To avoid fatigue, it is recommended that no more than four different attributes (wines) per day could be evaluated and to take short breaks between them (*see* **Note 9**).

3.3.1 Time-Intensity Curve Parameters

The data collected in the TI sensory analysis methodology are typically represented in the form of TI curves (Fig. 3). Data acquired from TI assessments correspond to intensity values for the attribute of interest at each time point over the time setting (Fig. 3). Therefore, the curves show the changes in the intensity of each sensory attribute during the consumption of the wine (swallowed or after expectoration), where the. X-axis is the time in seconds and the Y-axis represents the perceived intensity (at each second). In this way, these curves allow a detailed visualization of how the sensory characteristics of the product change over time and can be used to identify patterns and trends in the temporal evolution of sensory attributes. TI curves usually follow a similar pattern, starting with a rising phase, continuing with a stationary phase, and ending with a declining stage of perception. According to most TI analyses, the average curve of all (trained) panelists' responses minimizes individual panelist differences and visualizes product differences [10, 19–21].

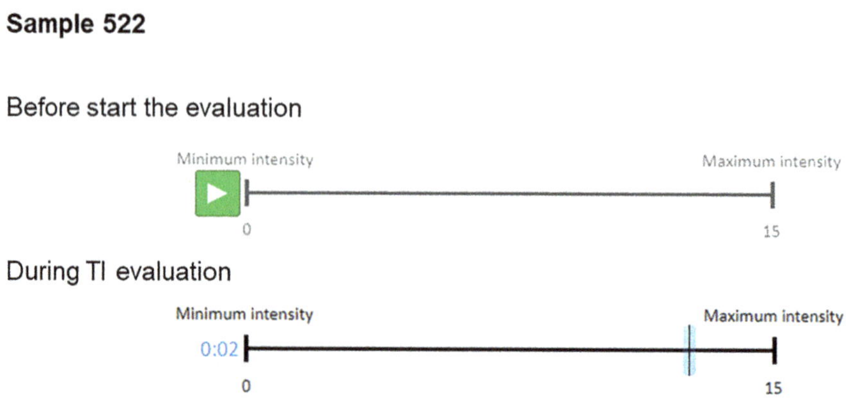

Fig. 2 Example of a screenshot corresponding to the evaluation of a flavor attribute using the TI methodology using the Compusense software

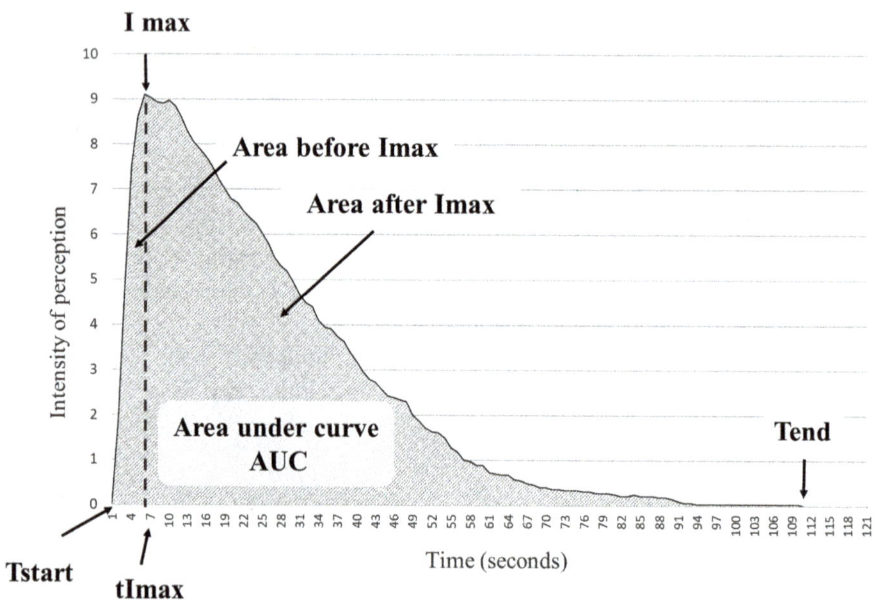

Fig. 3 Example of a Time-Intensity curve obtained by using the Compusense software, in which the different TI parameters are indicated

The most used parameters derived from the TI curves, which are used to compare among samples and that are detailed in Fig. 3 are: the time when the stimulus is first perceived after consuming the wine (T start), the maximum perceived intensity of the attribute (Imax), the time necessary to perceive the maximum intensity (tImax), the final time at which the attribute is no longer perceived (T end), the rate of intensity increase between T start and tImax (R increase); the rate of the intensity decrease between tImax and T end; and the area under the curve (AUC), which is related to the overall intensity of aroma perceived during the assessment [14].

The use of software simplifies the preparation of TI evaluation sessions. Nevertheless, there are parameters that need to be optimized for each analysis. The most important parameter is the duration of the analysis, which must be longer than the perception time in order to evaluate the final perception time (Tend) without bias. It must be considered that in some software (e.g., Compusense) the evaluation time does not end until the indicated time is over (*see* **Note 10**).

3.4 Statistical Data Analyses

Most sensory software allows to obtain the TI parameters from raw data. The analysis of the data typically involves generating an average TI curve and conducting ANOVA for each measured curve parameter to identify variations among products and assessors [18, 22]. In addition, a principal component analysis (PCA), polynomial and ordinary differential equation plot have also been employed to evaluated TI data [23, 24].

4 Notes

1. The panelists should be informed before the sensory sessions that no coffee, smoking or food is allowed 1 h before the tasting sessions.
2. It is important to check that the panel does not present olfactory or gustatory disorders such as anosmia, hyposmia, and dysgeusia.
3. After mouth rinsing with the sample, it is recommended to expectorate the sample to avoid increasing the fatigue of the panelist. Therefore, all panelists shall expectorate the sample to follow the same procedure.
4. The panel training should be conducted in several sessions on different days. Before starting the evaluation of the samples, it is necessary to ensure that the panel is trained.
5. It is advisable to train the panel in the use of the computer or tablet, as the computer skills of all participants are not equal and may induce bias in the results.
6. Some tools, such as TI-R and TI-RI [10, 25], can check the panel training by assessing intra-individual variation within the same panel. The lower the index, a better replication between curves are expected. Thus, these tools are able to detect which panelists are better trained or if there is a panelist who needs some more training since the panelists will be more constant throughout the training repetitions.

7. On each day of the session, it is recommended to do a warm-up assessment test, with a wine (different from the samples to be evaluated) to remind the panel how is the test procedure and the use of the software, therefore, to minimize the first position effect.
8. The panelists should not receive any information about his/her results of training and/or sensory evaluation.
9. Between samples, the panelists should drink water and/or eat palate cleaners such as bread crackers to clean the palate and reduce fatigue (1–2 min between samples). In the case of very strong or very astringent wines that make difficult to clean the palate with water and bread, they can rinse with water and pectin solutions (1 g/L) to avoid carry over between samples.
10. The analysis times set cannot be much longer than Tend, as the panelists would have to wait a long time to further evaluate the next sample.

Acknowledgments

This work has been done thanks to the financial support provided by AEI and MICIN (Project PID2019-11734-RBI00).

References

1. Muñoz-González C, Pozo-Bayón MÁ, Canon F (2021) Understanding the molecular basis of aroma persistence using real-time mass spectrometry. In: Dyn flavor capturing aroma using real-time mass spectrom, p 67–75
2. Lee WE III, Pangborn M (1986) Time-intensity: the temporal aspects of sensory perception. Food Technol 40:71–78
3. Duizer LM, Bloom K, Findlay CJ (1997) Dual-attribute time-intensity sensory evaluation: a new method for temporal measurement of sensory perceptions. Food Qual Prefer 8(4): 261–269
4. Kuesten C, Bi J, Feng Y (2013) Exploring taffy product consumption experiences using a multi-attribute time–intensity (MATI) method. Food Qual Prefer 30(2):260–273
5. Pineau N, Cordelle S, Schlich P (2003) Temporal dominance of sensations: a new technique to record several sensory attributes simultaneously over time. Journées de statistique 35:1
6. Castura JC, Antúnez L, Giménez A, Ares G (2016) Temporal Check-All-That-Apply (TCATA): a novel dynamic method for characterizing products. Food Qual Prefer 47:79–90
7. Cliff M, Heymann H (1993) Development and use of time-intensity methodology for sensory evaluation: a review. Food Res Int 26(5): 375–385
8. Gotow N, Omata T, Uchida M, Matsuzaki N, Takata S, Hagiwara I et al (2018) Multi-sip time–intensity evaluation of retronasal aroma after swallowing oolong tea beverage. Foods 7(11):177
9. Vilela A, Bacelar E, Pinto T, Anjos R, Correia E, Gonçalves B et al (2019) Beverage and food fragrance biotechnology, novel applications, sensory and sensor techniques: an overview. Foods 8(12):643
10. Chaya C, Criado C, Pozo-Bayón MÁ, Echevarrías-Marco A, Dupas de Matos A (2019) A new index for predicting differences in repeatability of time-intensity curves: time-intensity reliability index (TI-RI). Food Qual Prefer 76 (February):33–38. https://doi.org/10.1016/j.foodqual.2019.03.013
11. Visalli M, Galmarini MV (2022) Multi-attribute temporal descriptive methods in sensory analysis applied in food science: protocol for a scoping review. PLoS One 17(7): e0270969

12. Criado C, Muñoz-González C, Pozo-Bayón MÁ (2021) Differences in salivary flow and composition between age groups are correlated to dynamic retronasal aroma perception during wine consumption. Food Qual Prefer 87: 104046. https://doi.org/10.1016/j.foodqual.2020.104046
13. Peyvieux C, Dijksterhuis G (2001) Training a sensory panel for TI: a case study. Food Qual Prefer 12(1):19–28
14. Keefer HRM, Rovai D, Drake M (2023) A timely application – temporal methods, past, present, and future. J Food Sci 88(S1):A21–A52
15. Sokolowsky M, Fischer U (2012) Evaluation of bitterness in white wine applying descriptive analysis, time-intensity analysis, and temporal dominance of sensations analysis. Anal Chim Acta 732:46–52
16. Pérez-Jiménez M, Pozo-Bayón MÁ (2019) Development of an in-mouth headspace sorptive extraction method (HSSE) for oral aroma monitoring and application to wines of different chemical composition. Food Res Int 121: 97–107. https://doi.org/10.1016/j.foodres.2019.03.030
17. Gotow N, Moritani A, Hayakawa Y, Akutagawa A, Hashimoto H, Kobayakawa T (2015) Development of a time–intensity evaluation system for consumers: measuring bitterness and retronasal aroma of coffee beverages in 106 untrained panelists. J Food Sci 80(6): S1343–S1351
18. Lawless HT, Heymann H (2010) Sensory evaluation of food: principles and practices, vol 2. Springer, New York
19. Eilers PHC, Dijksterhuis GB (2004) A parametric model for time–intensity curves. Food Qual Prefer 15(3):239–245
20. Overbosch P, van den Enden JC, Keur BM (1986) An improved method for measuring perceived intensity/time relationships in human taste and smell. Chem Senses 11(3): 331–338
21. Dijksterhuis G, Eilers P (1997) Modelling time-intensity curves using prototype curves. Food Qual Prefer 8(2):131–140
22. Marques C, Correia E, Dinis L-T, Vilela A (2022) An overview of sensory characterization techniques: from classical descriptive analysis to the emergence of novel profiling methods. Foods 11(3):255
23. van Buuren S (1992) Analyzing time-intensity responses in sensory evaluation. Food Technol 46(2):101–104
24. Wendin K, Janestad H, Hall G (2003) Modelling and analysis of dynamic sensory data. Food Qual Prefer 14(8):663–671
25. Bloom K, Duizer LM, Findlay CJ (1995) An objective numerical method of assessing the reliability of time-intensity panelists. J Sens Stud 10(3):285–294

Chapter 15

Temporal Dominance of Sensations (TDS) Applied to Wine Sensory Evaluation

Alice Vilela

Abstract

Temporal Dominance of Sensations (TDS) is a temporal sensory methodology that consists in identifying and rating sensations perceived as dominant until the perception ends. This method does not require long training and enables the evaluation of several attributes at the same time. This method has been used to describe the temporality of sensations in wines.

Key words TDS curves, Dynamic wine sensory analysis, SensoMaker®, Sensory attributes, DR max, T max, T 90% max

1 Introduction

Wine sensory profile characterization is a recurrent theme, and descriptions of wines' sensory attributes, usually generated by wine specialists, are widely used to guide consumer purchases. Formerly, descriptive analysis was based on the methods of Quantitative Descriptive Analysis (QDA®), used worldwide, and considered a reference for reliable, and valid sensory analyses aiming to measure sensory properties among a set of samples [1]. However, the way consumers perceive wine is strongly linked to their expectations, based on their enjoyment and satiety. So, a temporal driver approach can trigger those expectations in consumers during the tasting process [2]. Over the last few years, temporal dynamics in the sensory assessment of wines have been broadly investigated [3, 4] since sensory evaluation is a complex and dynamic process that floats and evolves.

One of the first and currently most used temporal methodologies is Time–Intensity (TI), which consists in recording one by one the evolution, in terms of intensity, of given attributes. According to Pineau et al. [5], TI is a time-consuming method that has to be carried out with a limited number of attributes, since only one

María Ángeles Pozo-Bayón and Carolina Muñoz González (eds.), *Wine Analysis and Testing Techniques*, Methods and Protocols in Food Science, https://doi.org/10.1007/978-1-0716-3650-3_15,

attribute can be evaluated at a time. Besides, the continuous calculation of temporal changes in the perception of a single attribute is known to induce a halo-dumping effect [6] with a carryover from perceived attributes to the next being evaluated. To overcome these weaknesses, Pineau et al. [5] developed a new method called Temporal Dominance of Sensations (TDS), consisting in identifying and rating sensations perceived as dominant until the perception ends. This method does not require long training and allows the evaluation of several attributes at the same time. Pessina [7] was the first author to use this method to describe the temporality of sensations in wines. Since then, several authors have used this technique to evaluate and discriminate wines from different sub-regions of the same region [8], to evaluate the Influence of Wine on Cheese Perception [9], and even to analyze the Impact of music on the perception of red wine via Temporal Dominance of Sensations [10].

TDS can be performed using laptops and tablets and using different software such as XLSTAT SENSORY (by Lumivero) or the free SensoMaker® software (version 1.91, 2017, Universidade Federal de Lavras UFLA, Lavras, MG, Brazil) for data acquisition and analysis. The SensoMaker® software output shows the Temporal Dominance of Sensation curves (Fig. 1), along with some quantitative parameters of TDS curves (DR max—Highest maximum dominance rate; T max—Time for highest maximum dominance rate, and T 90% max—Time interval which dominance rate is ≥90% of DR max) (Table 1).

This chapter will provide a detailed explanation of how to perform a TDS evaluation on wine samples, and how to use the SensoMaker® software, data acquisition, and analysis.

The protocol described in this chapter is divided into two activities: First activity (activity 1) allows familiarizing with the software and is a simple procedure, with a three attributes evaluation in a glass of wine. Using this software, a minimum of one (1) and a maximum of eight (8) attributes may be evaluated. In the second activity (activity 2), eight aromatic attributes can be evaluated, in different red wines, as an example, once the selected descriptors are usually present in red wines.

Wines and attributes can be changed according to tasting objectives.

2 Materials

2.1 Material Per Panelist

1. Tasting glasses, one per wine, with 50 mL of wine to be evaluated, coded with random three-digit codes.
2. Mineral water and dried unsalted biscuits to clean the palate and taste buds between the tasting of each wine.

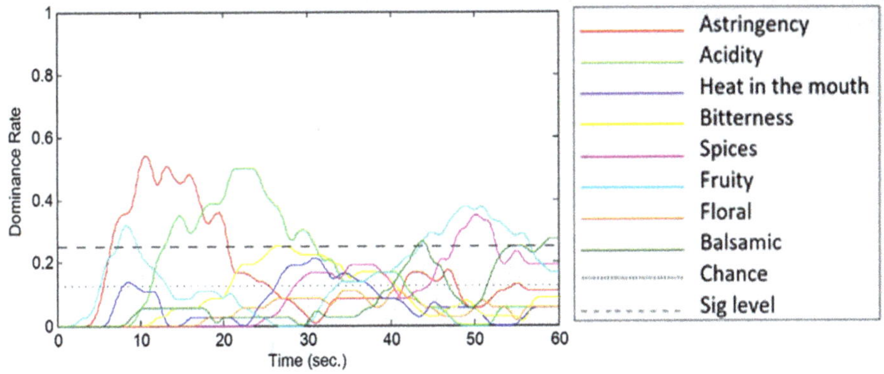

Fig. 1 TDS curves of taste/flavor attributes of red wines evaluated by an expert panel. For each attribute, a colored line is presented in each graphic. Lines of chance and significance level are also shown. The "chance level" represents the dominance rate that an attribute can obtain by chance (1/number of attributes), and the "significance level line" is created by a binomial test, which expresses the smallest value of the proportion being significantly higher than the chance level [5, 8]

Table 1
Quantitative parameters of the TDS curves[a]

Attributes								
Param	**Astringency**	**Acidity**	**Heat**	**Bitterness**	**Spices**	**Fruity**	**Floral**	**Balsamic**
DR max	54.29%	50%	21.46%	25%	35.10%	38.13%	11.11%	27.78%
T max	10.6 s	21.5 s	31 s	26.5 s	50.2 s	49 s	34.5 s	59.6 s
T 90% max	3.8 s	4 s	3.6 s	3.8 s	2.8 s	4.5 s	7.4 s	16.9 s

[a]s—time in seconds

3. Tasting glasses lids.
4. Spatters and napkins.

2.2 Laptops and Software

1. Prepare laptops with the selected software. The free SensoMaker® software (version 1.91, 2017, Universidade Federal de Lavras UFLA, Lavras, MG, Brazil) download from https://www.ufla.br/sensomaker/ has been chosen as an example to explain a general TDS procedure for wine TDS analysis.
2. Considering that SensoMaker® is a MATLAB stand-alone application. To run it, you must install MATLAB Compiler Runtime (MCR), which is freely available on the same web page.

3 Methods

3.1 Laboratory Conditions and Wines

1. Prepare the laboratory with all ISO 8589 [11] regulatory requirements, giving the panel an appropriate environment (*see* **Note 1**); such conditions are critical to ensure the quality and reliability of the results.
2. The wines must be enveloped and coded with random three-digit codes so as not to identify the labels (*see* **Note 2**), and opened 30 min in advance of the tasting.
3. Fifty milliliters (50 mL) of each tasting sample or wine must be served, at room temperature (20 ± 2 °C), in ISO 3591 [12] tasting glasses, coded, and randomly arranged.

3.2 Activity 1 (Familiarizing with the Software)

Open the software and choose the correct test to perform (Fig. 2);

3.2.1 Setting the Experiment

1. Set the instructions for the test using the menu File > Set instructions (for example, "When tasting the wine sample check the dominant sensation"), Fig. 3a.
2. Start with a simple tasting procedure by analyzing only three dominant sensations (sweet, sour, bitter), in one of the wine samples (*see* **Note 3**).
   (a) Select the directory to save (A).
   (b) Set the total time for evaluation (B).
   (c) If necessary, set the delay time (C) to be counted before starting the analysis (*see* **Note 4**).

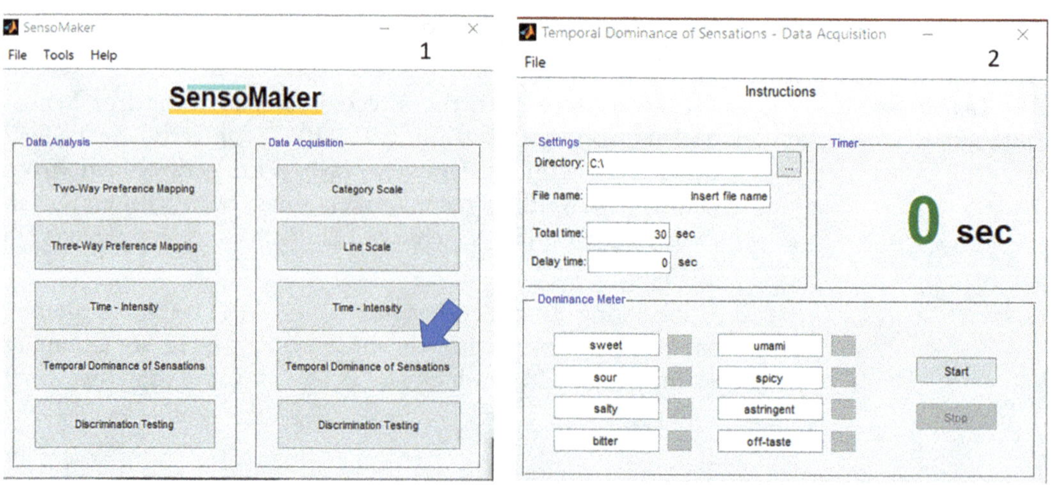

Fig. 2 SensoMaker's home screen with data acquisition bottoms on the right and data analysis on the left. For performing a TDS test press Temporal Dominance of Sensations (blue arrow)

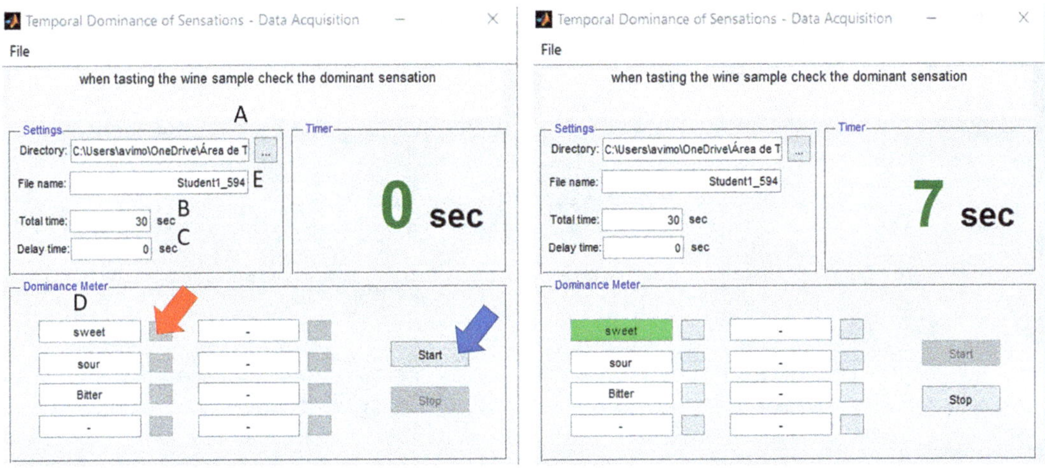

Fig. 3 Setting (**a**) and performing (**b**) the experiment

(d) Set the sensations/attributes names (D) and set a dash (−) for not used sensations. Leaving it blank is not appropriate.

3.2.2 Performing the Experiment

1. Insert the file name (E). In this case (example above), it is the student name and the sample code (Student1_594).
2. Try the sample and press the Start button (blue arrow).
3. Check the dominant sensation (D) using the proper button. The chosen sensation turns green (Fig. 3b).
4. When completed, a successful message is shown and then the window is ready for a new analysis (analysis of a new sample).
5. You can stop the analysis by pressing the STOP button.
6. All the analyses must be stored in the same directory (folder) (*see* **Note 5**).

3.2.3 Analyzing the Data

1. On the Temporal Dominance of Sensations data analysis module (Fig. 4), press Import Data (A).
2. Select all files (files from the class students, all in the same folder) obtained after evaluation of the selected wine sample and press Open.
3. If appropriate, disabled sensations not be analyzed (B) (*see* **Note 6**).
4. Set the smooth level for the curve (C). If smooth is not appropriate, disable this option (D).
5. Chance level (E) and significance level (F) lines also can be disabled.
6. Press the Plot button (G) to obtain the curves (Fig. 4, Data 1, and Data 2).

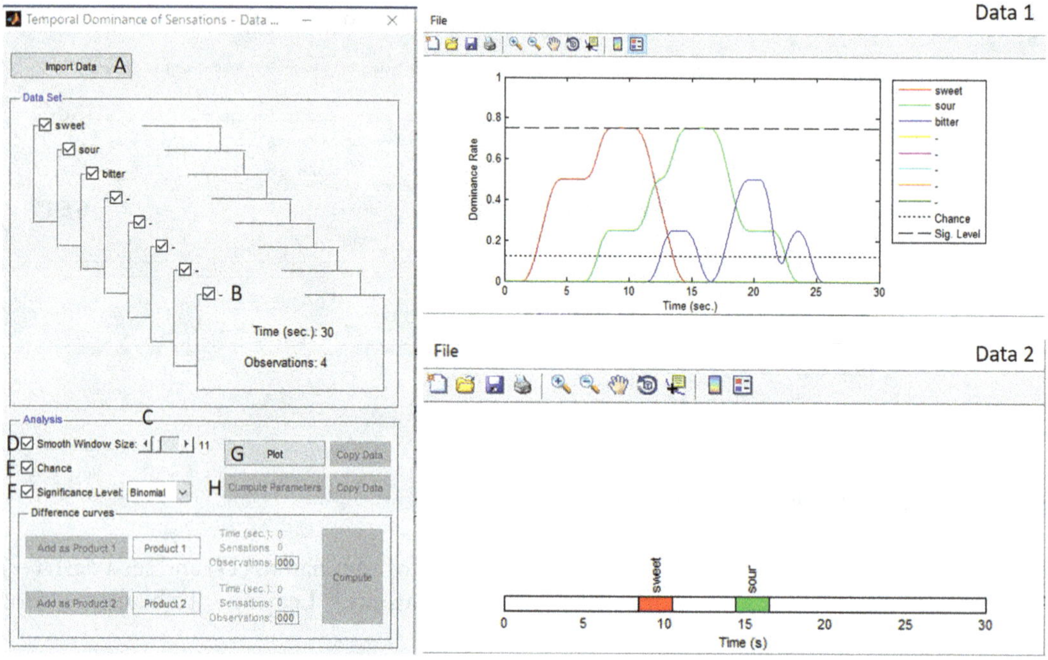

Fig. 4 Analyzing Temporal Dominance of Sensations tests and Curve parameters (Data 1) for sweet, sour, and bitter sensations of example wine 594. Data 2 shows the time, in seconds, in which sweet (red) and sour (green) sensations were at a significant level. Note that the bitter sensation does not reach a significant level

Table 2
Quantitative parameters of the TDS curves for sweet, sour, and bitter sensations of example wine 594[a]

Attributes			
Bitter	**Sweet**	**Sour**	
DR max	75%	75%	50%
T max	8.5 s	14.5 s	19.5 s
T 90% max	3.4 s	3.4 s	2.0 s

[a]s—time in seconds. DR max—highest maximum dominance rate; T max—time for highest maximum dominance rate, and T 90% max—Time interval in which dominance rate is ≥90% of DR max

7. Press Compute Parameters button (Fig. 4h) to obtain quantitative curve parameters (Table 2).

Analyzing Table 2, the sweet and sour tastes' highest maximum dominance rate (DR max) is 75%, while for bitter taste, it is only 50%; the time for the highest maximum dominance rate (T max) for sweet-taste occurred at 8.5 s, for sour taste at 14.5 s and for bitter at 19.5 s of analysis. The time interval in which the dominance rate is ≥90% of DR max (T 90% max) lasted 3.5 s for sweet and sour, and 2.0 s for bitter taste.

3.3 Activity 2 (Multiple Attributes Evaluation)

Usually, if more than 3 attributes are to be evaluated, a TDS protocol (Table 3) must be followed so that the evaluation procedure is similar among panelists (*see* **Note** 7). In Table 3, an example of a TDS assessment protocol, adapted from Correia et al. [8], is shown.

3.3.1 TDS Wines Olfactory Assessment

Begin by setting the experiment as explained in Subheading 3.2.1, and choose the attributes you want to evaluate (*see* **Note 8**). Do not forget that they must be olfactory attributes.

Set the delay time (Subheading 4, **step 1**, Fig. 3c), which in the example protocol is 5′ (five seconds). In these 5′ you are in **steps 1** and **2** of the protocol (Table 3). You are preparing the wine for evaluation (shaking the tasting glass clockwise, leading to the release of wine aromas to the interior of the glass).

In **step 3**, the sensory evaluation begins. When the clock marks 0 (zero), smell the glass continuously for 8′ and at the same time click on one of the listed attributes that correspond to the most dominant at the moment. Click on a new attribute whenever you feel dominance change. You may click as many times as you need.

At **step 4**, from 9′ to 11′, you must clean the air of your nose so that you may perceive more aromas (*see* **Note 9**). Distance the tasting glass from the nose; inhale and exhale for 2′; then re-smell the tasting glass to continue the assignment of dominance by 7′.

In **step 5**, from 18′ to 22′, Shake the glass clockwise for 4′. Then re-smell the glass to continue the assignment of dominance by 7′.

In **step 6**, from 29′ to 31′, repeat **step 4**. Distance the tasting glass from the nose; inhale and exhale for 2′; then re-smell the tasting glass to continue the assignment of dominance by 7′.

In **steps** 7 and **8**, repeat **step 5** and then **step 4** again.

In **step 9** (at the end of 60′) you end your evaluation.

Note that this is an evaluation for 60 s. If the evaluation is longer, you can use this protocol and repeat **steps 5** and **4** as many times as you need.

3.3.2 TDS Wines Taste/Flavor Assessment

Begin by setting the experiment as explained in Subheading 3.2.1, and choose the attributes you want to evaluate (*see* **Note 8**, again). Do not forget that they must be taste/flavor attributes.

Set the delay time (Subheading 4, **step 1**, Fig. 3c), which in the example protocol is 5′ (five seconds). In these 5'you are in **steps 1** and **2** of the protocol. You are preparing the wine for evaluation. Remove the lid from the tasting glass and hold it with your left hand.

With the right hand start the evaluation by clicking with the cursor in Start; bring the wine to the mouth and make it evenly distributed, then discard. Do not evaluate this first contact with the wine (*see* **Note 10**).

Table 3
TDS assessment example protocol for olfactory and taste/flavor attributes

Olfactory assessment			Taste/flavor assessment		
Step	Time (s)	Instructions	Step	Time (s)	Instructions
1	–	Remove the lid from the tasting glass and hold it with your left hand	1	–	Remove the lid from the tasting glass and hold it with your left hand
2	−5′ (delay time)	With the right hand start the evaluation by clicking the cursor on the start button and shaking the tasting glass clockwise for 4′. If you have difficulty shaking the glass, use the table as a support base	2	−5′ (delay time)	With the right hand start the evaluation by clicking with the cursor in start; bring the wine to the mouth and make it evenly distributed, then discard. Do not evaluate this first contact with the wine
3	0′	Smell the glass continuously for 8′ and at the same time click on one of the listed attributes that correspond to the most dominant at the moment. Click on a new attribute whenever you feel dominance change	3	0′	Take the wine to the mouth and keep it for 4′; have it distributed evenly and at the same time click on one of the listed attributes that match the most dominant at the moment. Click on a new attribute whenever you feel dominance change. Wine can be swallowed or discarded
4	From 9′ to 11′	Distance the tasting glass from the nose; inhale and exhale for 2′; then re-smell the tasting glass to continue the assignment of dominance by 7′	4	From 5′ to 14′	Continue the evaluation and attribution of the dominant sensations by 9′
5	From 18′ to 22′	Shake the glass clockwise for 4′. Then re-smell the glass to continue the assignment of dominance by 8′	5	From 15′ to 19′	Repeat step 3
6	From 29′ to 31′	Repeat step 4	6	From 20′ to 29′	Repeat step 4
7	From 38′ to 42′	Shake the glass clockwise for 4′. Then re-smell the tasting glass to continue the assignment of dominance by 6′.	7	From 30′ to 34′	Repeat step 3
8	From 48′ to 52′	Distance the tasting glass from the nose; inhale and exhale for 2′; then re-smell the tasting glass to continue the assignment of dominance by 8′	8	From 35′ to 44′	Repeat step 4

(continued)

Table 3 (continued)

Olfactory assessment			Taste/flavor assessment		
Step	**Time (s)**	**Instructions**	**Step**	**Time (s)**	**Instructions**
9	60′	End of evaluation	9	From 45′ to 49′	Repeat step 3
			10	From 50′ to 59′	Repeat step 4
			11	60′	End of evaluation

Adapted from Correia et al. [8]

In **step 3**, the sensory evaluation begins. When the clock marks 0 (zero), take the wine to the mouth and keep it for 4′; have it distributed evenly and at the same time click on one of the listed attributes that match the most dominant at the moment. Click on a new attribute whenever you feel dominance change. Wine can be swallowed or discarded.

At **step 4**, from 5′ to 14′, continue the evaluation and attribution of the dominant sensations by 9′.

In the following steps, you will repeat **steps 3** and **4**, as many times as you need, until the end of the evaluation.

3.4 Expected Outcomes and Results Analysis

For obtaining and analyzing the TDS curves prosed as explained in Subheading 3.2.3. As is possible to see in the example (Fig. 5), for each evaluated wine and each time of evaluation, dominant rates, and curves are plotted by attribute. The dominant rates are obtained by dividing the number of citations of an attribute by the number of panelists and the number of replications. Since one panelist can have only a single dominant attribute at each time, the sum of the dominance rates over attributes is equal to one at each time; the higher the dominant index, the better the agreement among panelists.

The graphics represent two other lines: the "chance level" which represents the dominance rate that an attribute can obtain by chance (1/number of attributes), and the "significance level line", based on a binomial test, which expresses the smallest value of the proportion being significantly higher than the chance level. When the TDS curves go from between the chance and the significance levels to above the latter, they are consistent at the panel level.

In the olfactory evaluation of wines from A, according to the graphic representation of the TDS curves (Fig. 5), six attributes overlapped with the significance level line, according to the perception of the tasters: Floral, Fresh Fruit, Spices, Balsamic, Ripe Fruit, and Empyreumatic.

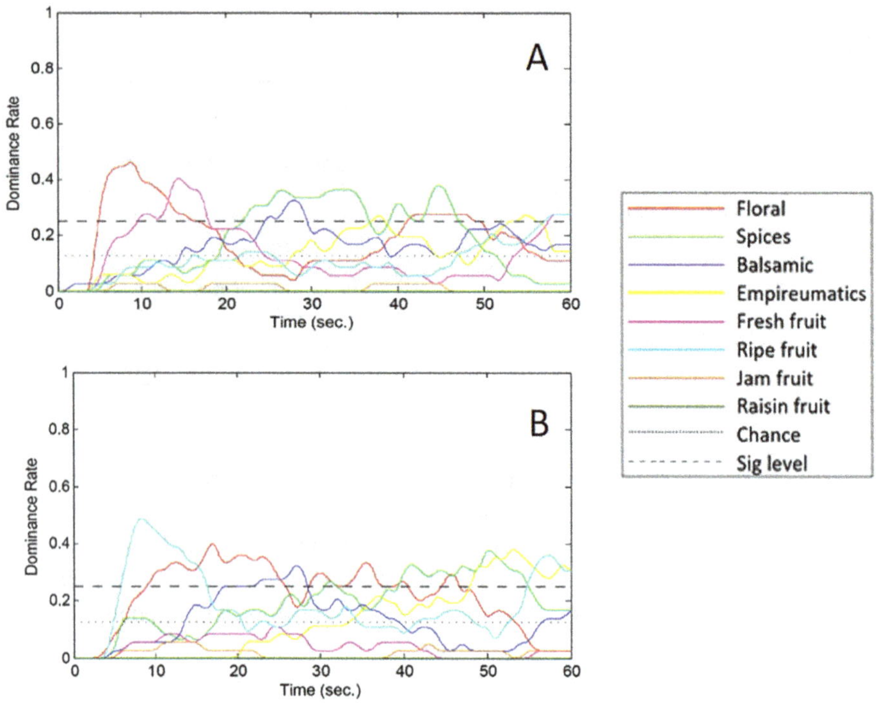

Fig. 5 TDS curves of olfactory attributes of wines from two wine regions (**a** and **b**, six wines per region). For each attribute, a colored line is presented in each graphic. (Adapted from Correia et al. [8])

Analyzing Table 4, quantitative parameters of TDS curves, it is possible to verify that the Floral attribute recorded the highest maximum dominance rate (DR max) with the representativeness of 46.09% of the evaluations, at 8.7 s (seconds) of the test, followed by attributes Fresh fruit (DR max 40.45%) at 14.3 s, Spices (DR max 37.88%) at 44.8 s, Balsamic (DR max 32.58%) at 28 s, Ripe fruit (DR max 27.78%) at 58.5 s and Empyreumatic (DR max 27.02%) at 55 s. The Spices attribute stood out with the largest range of the maximum dominance rate (T 90% max) lasting 20.1 s.

In the olfactory evaluation of wine B according to the interpretation of the TDS curves, the attributes: Ripe Fruit, Floral, Balsamic, Spices, and Empyreumatic, overlapped the line of significance level and were ordered according to the perception of the tasters.

Analyzing the quantitative parameters of TDS curves, Table 4, The highest maximum dominance rate was obtained by the attribute Ripe fruit (DR max 48.86%) at 8.3 s of the evaluation, successively by Floral (DR max 38.95%) at 16.9 s, Empyreumatic (DR max 37.88%) at 53.2 s, Spices (DR max 37.75%) at 50.3 s, and Balsamic (DR max 32.32%) at 27.2 s. The Floral attribute was expressed with the longest time interval of the dominance rate (T 90% max) with a duration of 4.8 s.

Table 4
Quantitative parameters of the TDS curves A and B[a]

Attributes									
Wines		**Floral**	**Spices**	**Balsamic**	**Empyreumatic**	**Fresh fruit**	**Ripe fruit**	**Jam fruit**	**Raisin fruit**
A	DR max	46.09%	37.88%	32.58%	27.02%	40.45%	27.78%	2.78%	0%
	T max	8.7 s	44.8 s	28 s	55 s	14.3 s	58.5 s	7.5 s	0 s
	T 90% max	3.5 s	20.1 s	2.1 s	20 s	2.4 s	2.9 s	37.8 s	0 s
B	DR max	38.95%	37.75%	32.32%	37.88%	11.11%	48.86%	5.56%	0%
	T max	16.9 s	50.3 s	27.2 s	53.2 s	24.5 s	8.3 s	11.5 s	0 s
	T 90% max	4.8 s	2.2 s	1.8 s	4 s	1.4 s	2.8 s	4.2 s	0 s

Adapted from Correia et al. [8]

[a]s—time in seconds. DR max—highest maximum dominance rate; T max—time for highest maximum dominance rate, and T 90% max—Time interval in which dominance rate is ≥90% of DR max

Data analysis can be performed by Multivariate analysis of variance (MANOVA) to evaluate the significance of the samples on the quantitative parameters of TDS curves. When MANOVA detects statistically significant effects, univariate analysis of variance (ANOVA) may be performed, followed, whenever possible, by Tukey's post-hoc test [8].

4 Notes

1. Wine sensory evaluation is the process of analyzing and evaluating the aromas, flavors, and other sensory characteristics of a wine. To perform this evaluation effectively, it is important to create an appropriate environment that allows for accurate and consistent sensory analysis. Here are some factors to consider when creating a suitable environment for wine sensory evaluation: (a) The lighting in the room should be bright enough to allow for accurate color evaluation of the wine, but not so bright that it affects the taster's ability to evaluate the aroma and flavor of the wine. Natural light is ideal, but if this is not possible, use white light bulbs with a color temperature of around 5500 K; (b) Temperature and humidity: The room should be maintained at a temperature between 18 and 22 °C and a humidity level of around 60%. This will ensure that the wine is not affected by extreme temperatures or humidity levels, which can alter its sensory characteristics; (c) The room should be clean and free from any strong odors or aromas that could interfere with the evaluation of the wine. Avoid wearing perfume or cologne, and ensure that the room is well-ventilated; (d) Glassware: Wine is best evaluated in a

standardized glass that allows for proper aeration and allows the taster to evaluate the color, aroma, and flavor of the wine. Use clear, odor-free wine glasses with a tulip-shaped bowl that narrows towards the top to concentrate the aromas; and (e) Neutral palate cleansers: Provide neutral palate cleansers, such as crackers or bread, and room temperature water to help tasters cleanse their palate between tastings.

2. Wines must be enveloped and coded with random three-digit codes when performing sensory evaluation tests to eliminate potential biases that can arise from preconceived notions or expectations of the wine based on the label or branding. For example, if a taster sees a well-known and respected brand on the label, they may expect the wine to be of higher quality and rate it accordingly, even if the wine does not meet their actual sensory standards. Alternatively, if a taster sees a label from a lesser-known or inexpensive brand, they may expect the wine to be of lower quality and rate it lower, even if the wine actually meets their sensory standards. By enveloping and coding the wines with random three-digit codes, the tasters are forced to evaluate the wines solely based on their sensory characteristics, without any preconceived notions or biases based on the label or branding. This allows for a more objective and accurate evaluation of the wine's sensory qualities.
3. Wine can have a variety of dominant sensory sensations, depending on the type of wine and the specific characteristics of that wine. For example, red wines are typically associated with dominant sensory sensations such as tannins, acidity, and fruit flavors. Acidity, which is the tart or sour taste in wine, can add to the wine's freshness and liveliness. Fruit flavors, such as blackberry, cherry, and plum, are common in red wines. White wines are typically associated with dominant sensory sensations such as acidity, minerality, and floral or citrus aromas. White wines tend to have higher acidity than red wines, which can give them a crisp, refreshing taste. Minerality, which can come from the soil in which the grapes are grown, can add a subtle, earthy flavor to the wine. Floral or citrus aromas, such as lemon, lime, or grapefruit, can make the wine smell fresh and bright.
4. If, before carrying out the evaluation, it is necessary, for example, to shake the glass to allow the aromas of the wine to be released, the timer can be adjusted allowing a few seconds before the start of the evaluation itself.
5. Each analysis generates a file. The files must be saved in a folder, with, for example, the name/number of the tasting session. Thus, when it is necessary to analyze the data, the files corresponding to the session can be selected from a specific folder.

6. Even if the test has more than one sensation/descriptor, during the analysis, sensations/descriptors can be selected, individually or in groups.
7. There are some constraints associated with TDS wine analysis: (a) Subjectivity: TDS analysis relies on the subjective interpretation of sensory attributes by the taster. The results of the analysis can vary depending on the individual taster's sensory sensitivity and personal preferences; (b) Complexity: TDS analysis can be a complex and time-consuming process, requiring skilled tasters and specialized equipment to accurately record and analyze the data. The complexity of the analysis can make it difficult to implement on a large scale or in certain settings; and (c) Standardization: There is a lack of standardization in TDS analysis, with different studies using different protocols and parameters for data collection and analysis. This can make it difficult to compare results across studies or to establish best practices for the technique. So, to minimize these constraints, developing a protocol to be used and implemented in the panel is appropriated.
8. For the correct selection of attributes, a free profiling test (FP) can be carried out, where tasters are asked to freely choose attributes that best describe the wine. Afterward, those who present a citation percentage greater than 50% (or the one that the panel leader deems advisable) are selected. The chosen descriptors can then be applied in TDS tests or other single-point descriptive tests such as Quantitative Descriptive Analysis (QDA) or a simple CATA (Check-All-That-Apply) test.
9. It is important to clean the air of your nose during a TDS evaluation to ensure that you are able to accurately perceive and distinguish between different sensory attributes over time. TDS method requires the taster to pay close attention to the dynamic changes in the sensory attributes and to differentiate between them accurately. The olfactory system, which is responsible for detecting aromas, is an essential component of TDS evaluation. If the nasal passages are blocked or congested, the flow of air through the nose is impeded, and the ability to perceive aromas is reduced. This can significantly impact the taster's ability to accurately evaluate the sensory attributes of the wine over time. By cleaning the air of your nose, you can remove any irritants that may be affecting your ability to smell properly. This can help to activate the sensory receptors in the nose and allow you to accurately perceive and distinguish between different sensory attributes over time. This is essential for obtaining accurate and reliable TDS evaluation results.

10. In wine sensory evaluation, the first mouth contact with the wine is often not counted for the evaluation because it is considered a prelude to the main tasting experience. During the first sip, the mouth is not yet acclimated to the wine, and the initial shock of the taste and texture can be overwhelming. This can make it difficult to accurately evaluate the wine's sensory attributes and to differentiate between them. Instead, tasters will typically take a small sip of wine and hold it in their mouth for a few seconds before swallowing or spitting it out. This allows the mouth to become acclimated to the wine and for the taster to better evaluate the wine's sensory attributes, such as its flavor, texture, and finish. In some cases, the first sip of wine may be evaluated separately, as it can provide information about the initial impression of the wine and any sensory attributes that are immediately noticeable. However, it is generally not considered as part of the main evaluation process, as it may not accurately reflect the overall quality and character of the wine.

Acknowledgments

The author would like to thank the Chemistry Research Center (CQ-VR) and the e-Flavor Project for their financial support.

Funding This work was funded by the Chemistry Research Center (CQ-VR), (grant number UIDB/00616/2020 and UIDP/00616/2020), and by the e-Flavor Project, operation POCI-01-0247-FEDER-049337 financed by the FEEI and FEDER, under the Competitiveness and Internationalization Operational Program.

References

1. Marques C, Correia E, Dinis L-T, Vilela A (2022) An overview of sensory characterization techniques: from classical descriptive analysis to the emergence of novel profiling methods. Foods 11:255
2. Nguyen QC, Varela P (2021) Identifying temporal drivers of liking and satiation based on temporal sensory descriptions and consumer ratings. Food Qual Prefer 89:104143
3. Pierguidi L, Spinelli S, Monteleone E, Dinnella C (2021) The combined use of temporal dominance of sensations (TDS) and discrete time-intensity (DTI) to describe the dynamic sensory profile of alcoholic cocktails. Food Qual Prefer 93:104281
4. Silva AP, Voss HP, van Zyl H, Hogg T, de Graaf C, Pintado M, Jager G (2019) Effect of adding hop aroma in beer analysed by temporal dominance of sensations and emotions coupled with temporal liking. Food Qual Prefer 75:54–63
5. Pineau N, Schlich P, Cordelle S, Mathonnière C, Issanchou S, Imbert A et al (2009) Temporal dominance of sensations: construction of the TDS curves and comparison with time-intensity. Food Qual Pref 20(6): 450–455
6. Clark CC, Lawless HT (1994) Limiting response alternatives in time-intensity scaling:

an examination of the halo-dumping effect. Chem Senses 19(6):583–594

7. Pessina R. (2006) Temporal dominance of taste and flavour of wine (PhD thesis). ENSBANA, Dijon/Università Degli Studi, Foggia, Italy. Available online: http://www.sudoc.fr/109217985. Accessed on 20 May 2019
8. Correia E, Amorim E, Vilela A (2022) Structural equation modeling (SEM) and temporal dominance of sensations (TDS) in the evaluation of DOC Douro red wine's sensory profile. Foods 11(8):1168
9. Galmarini MV, Loiseau AL, Debreyer D, Visalli M, Schlich P (2017) Use of multi-intake temporal dominance of sensations (TDS) to evaluate the influence of wine on cheese perception. J Food Sci 82(11):2669–2678
10. Wang QJ, Mesz B, Riera P, Trevisan M, Sigman M, Guha A, Spence C (2019) Analysing the impact of music on the perception of red wine via temporal dominance of sensations. Multisens Res 32(4–5):455–472
11. ISO 8589 (2007) Sensory analysis – general guidance for the design of test rooms; ISO/TC 34/SC 12 sensory analysis. International Organization for Standardization, Geneva
12. ISO 3591 (1977) Sensory analysis – apparatus – wine-tasting glass; ISO/TC 34/SC 12 Sensory Analysis. International Organization for Standardization, Geneva

Chapter 16

Evaluation of Hedonic and Emotional Response Evoked by Wines

Carolina Chaya and María Mora

Abstract

The study of the emotional response triggered by the consumption of foods and beverages has gained significant importance in the last decade. It has been shown that emotions are determinant in the decision-making process, and therefore their study could provide some clues to understand the key drivers of consumers' choices and preferences. This chapter aims to explain in detail the practical considerations that researchers must take into account when conducting a study of hedonic and emotional responses evoked by wines.

Key words Sensory science, Consumer research, Consumer response

1 Introduction

Success on new product development depends on fulfilling expectations of consumers in terms of choices and preferences [1]. Over 75% of new products, belonging to food and beverages categories, fail during the first year after being launched to the market [2]. Therefore, industry needs to improve its strategies to gain a better understanding of consumer' preferences and choices.

Sensory science helps to understand consumers' perception of a product by studying its properties and the interaction with the human senses. In other words, to study the sensory attributes of products and consumers' preferences is essential for industry to find the key drivers for the optimal design and development of products [3].

Measuring the hedonic response elicited by a product is an appropriate method to determine the product acceptance, and therefore, it is used across scientific studies to define consumer preferences and predict potential success. Other studies relate hedonic response with the sensory and physicochemical properties of the product; this information has proved to be useful to

María Ángeles Pozo-Bayón and Carolina Muñoz González (eds.), *Wine Analysis and Testing Techniques*, Methods and Protocols in Food Science, https://doi.org/10.1007/978-1-0716-3650-3_16,

investigate the specific sensory characteristics, which drive consumers' liking. However, it has been argued that the hedonic response itself is not always enough to explain consumers' choices and attitudes to different food products [4–6]. For example, Ng et al. [5], in their study about blackcurrant squashes, showed that emotions could discriminate among products with similar hedonic scores. Also, the authors showed a relationship between the emotions experienced by consumers and the sensory properties of the products.

One of the aspects that differentiate consumers' responses toward food product consumption and choice, beyond liking, has been the emotional response. Damasio [7] showed that the decision-making process is completely influenced by the role of emotions. His work in the field of neuroscience showed that the emotions were not independent of rationality, both are inseparable and are interlinked in the decision-making process [7]. Emotions have been found to affect consumers' behavior. Moreover, different aspects of a product such as appearance, flavor, presentation, packaging, and/or consumption context, have an influence on the emotional response [8–11]. Thus, a deep knowledge about how to measure and understand emotions might be one of the clues to understand which of the extrinsic and/or intrinsic sensory properties are key drivers of preferences, providing a competitive advantage to food industry in the marketplace [12].

Authors such as King & Meiselman [13] have described "emotion" as "brief, intense and often focused on a referent" response to a stimulus; Sander [14] described it as an "event focused, two-step, fast process consisting of (1) relevance-based emotion elicitation mechanism that (2) shape a multiple emotional response." From a neurophysiology point of view, Damasio [7] described emotions as "bioregulatory reactions that aim at promoting, directly or indirectly, the sort of physiological states that secure not just survival, but survival regulated into the range that humans, conscious and thinking creatures, identify with well-being." Coppin and Sander [15] proposed that the multicomponent character of emotions and three additional criteria were appropriate to define what an emotion is. The three different criteria suggested were: (a) emotions are two-step processes composed by an emotion elicitation mechanism and a response mechanism, (b) objects or situations are necessary for emotions to occur, and (c) its duration is brief and has a quick onset. Therefore, according to the length, emotions could be distinguished from other affective states whose duration is longer, for example, preferences (liking/disliking), attitudes (predisposition toward specific products), or moods (low intensity and long duration feelings) [15].

Emotion measurement on beverages has been previously addressed by Van Zyl and Chaya [16]. The current chapter focuses on methodological contributions to the study of hedonic and emotional response evoked specifically by wine products. Implicit methods to measure the emotional response are out of the scope of the present chapter. Practical considerations about measuring explicit emotional response are included.

2 Materials

2.1 Samples and Containers

1. Selection of a neutral warm-up sample (*see* **Note 1**) to avoid the first position effect [17, 18].
2. Studied samples whether commercial or experimental wines.
3. Wine containers: selection of disposable materials to serve the different samples to consumers. It must be the same for all of the samples (including the warm-up one). An example of a container could be the normalized glass for wine-tasting stated by ISO (ISO 3591:1977).

2.2 Collecting Instruments

1. Electronic devices associated with data acquisition software to capture data such as Compusense (Compusense Inc., Canada), FIZZ (FIZZ Biosystemes, France), RedJade Sensory Software (RedJade Software Solutions, LLC), SENSESBIT (TasteLab, Spain), etc.
2. Paper forms.

2.3 Consumers' Panel: Selecting a Representative Sample of Consumers for the Study

1. Defining the recruitment criteria (*see* **Note 2**) [2].
2. Defining the consumers' sample according to the aim of the study and the available resources.
3. Defining consumers' sample segmentation (e.g., age, gender, education background, type of wine preferences, etc.).

2.4 Facilities: Defining the Type of Study

1. Central location test, sensory lab, real context study (e.g., restaurant), or immersive and virtual reality contexts.
2. Home use test.
3. Online test.

3 Methods

3.1 General Procedure to Conduct Consumer Studies

1. Balancing or randomizing the order of samples presentation removes several sources of error. This does not apply to the warm-up sample, which should be always served in the first position.
2. To avoid psychological errors that commonly occur in sensory science, 3-digit random codes should be used to label all the samples.

3. Careful consideration should be given to sample volumes served to the consumers, due to sensory fatigue and alcohol content of wines (*see* **Note 3**).
4. Palate cleansers to avoid carry-over effects and adaptation to sensory stimuli are required before tasting each sample. Water and neutral breadsticks or crackers are successful in cleansing palate.

3.2 Hedonic Tests to Measure Acceptance

1. Different methods have been developed to measure acceptability in the Sensory Science. The most used have traditionally been the measurement of preference, and the measurement of acceptance. In preference measurement, the consumer has a choice. One product is to be chosen over one or more other products. In the measurement of acceptance or liking the consumer rate their liking for the product on a scale. Acceptance measurements can be done on single products and do not require a comparison to another product (*see* **Note 4**) [3].

3.3 Emotions: Type of Questionnaires

1. Non-verbal self-reported measures: Non-verbal self-reported measures avoid the necessity for language, allowing for potential cross-cultural application. Considering the universality of facial expressions, different methods for measuring the emotional response elicited by food products, which minimize cultural differences interpretation, have been developed: the Self-Assessment Manikin (SAM) [19], Product Emotion Measurement Instrument (PrEmo®) [20] and different lists of emojis (e.g., emojis list developed for assessing wine [21]) (*see* **Note 5**).
2. Verbal self-reported measures: Verbal self-reported measures use lists of emotional words to assess the feelings triggered by an event or stimulus (*see* **Note 6**). Emotion questionnaires used in food product research usually consist of a list of feelings that could vary in the length and in the nature of used words. This variation depends on the specificity of the lexicon and the consumers' culture. There are general emotional lexicons to be applied in a wide variety of products (Table 1), and there are specific emotional lexicons developed for assessing the emotional response evoked by wines (Table 2) (*see* **Note 5**).
3. Evoking the consumption context of the product before starting the emotional questionnaire, is a good practice to facilitate the task to the consumers [28] (*see* **Note 7**).
4. It should be noted that the self-reported methods (verbal and non-verbal) are also influenced by the usual factors involved in any sensory method: intrinsic factors related to the product (package, price, information, etc.) and extrinsic factors (related

Table 1
Overview of generic emotional lexicons in the sensory science literature

References	Instrument	Number of terms	Consumers nationality
King and Meiselman [13]	EsSense	39	English (EEUU)
Thomson and Crocker [22]	Lexicon	59	English, French, German and Italian
Gmuer et al. [23]	Lexicon	49	German
Dorado et al. [17]	EsSense	39	Spanish (Spain)
Nestrud et al. [24]	EsSense25	25	English (EEUU)

Table 2
Overview of wine emotional lexicons in the sensory science literature

References	Instrument	Number of terms	Language
Ferrarini et al. [9]	Lexicon	16	Italian (Italy)
Danner et al. [25]	Australian Wine Evoked Emotions Lexicon (AWEEL)	19	English (Australia)
Silva et al. [26]	Lexicon	29	Dutch (Netherlands) Portuguese (Portugal)
Mora et al. [27]	Lexicon	Conventional: 13 categories Rapid method: 15 categories	Spanish (Spain)

to the consumer, such as physiological, psychological, and cultural factors) (*see* **Note 8**).

3.4 Data Nature: Scales to Apply on Emotional Questionnaires (Verbal and Non-verbal)

1. Linear scales use to be anchored from "very low" to "very high." These scales provide continuous data.
2. Discrete scales must have an odd number of categories also anchored from "very low" to "very high." They provide discrete data, although under some assumptions they may behave as continuous data (e.g., depending on the number of discrete points and the number of collected responses).
3. Nominal scales (e.g., Check-All-That-Apply data) show a list of emotional terms that could be selected (or not) by the consumer according with their presence (or absence). These data are also considered as qualitative or categorical.

3.5 Statistical Analysis

1. Independently of the type of method, verbal or non-verbal, the selection of the appropriate method of statistical analysis is related to the data nature (continuous, discrete, or categorical).

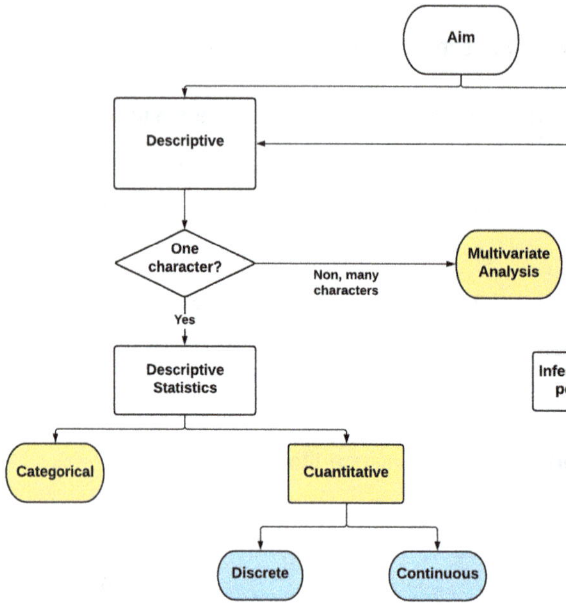

Fig. 1 Flowchart for choosing the most suitable statistical approach. (Source: Chaya [29])

A reflection must also be done on the descriptive vs inferential approach to analyze the data [29] (Fig. 1).

3.6 Ethical Committee Approval

1. Consumer studies should be conducted ensuring that the work described has been carried out in accordance with The Code of Ethics of the World Medical Association (Declaration of Helsinki). The protocol and procedures used in each study should be approved by an external ethical committee.
2. The experimental procedure must be explained and a written consent indicating voluntary participation need to be obtained from each participant prior to beginning the study.

4 Notes

1. Warm-up sample: also named dummy sample. It has been demonstrated that when several products are tested sequentially, consumers' response of the first-tasted product is overvalued. The introduction of a supplementary sample, whose data will be discarded, helps to avoid this effect. The sample should be as neutral as possible and it could belong, or not, to the set of assessed samples.
2. The recruitment criteria can vary depending on the objective of the study. The screening for selection typically involves a number of simple questions, for example, personal details, demographics, etc.

3. To avoid carry over effects by limiting the number of samples to taste by the consumer [30]. Authors recommend 25 mL per wine sample and not more than 5 or 6 samples in one session.
4. In preference tests, consumers are requested to rank samples from the least preferred to the most preferred. In the measurement of acceptance, the consumers are asked to rate their liking for the product using a scale. The 9-point hedonic scale is the most used scale ranging from dislike extremely to like extremely. The choice between both approaches has an effect on the statistical method of analysis.
5. Randomizing the order of the emotional terms/emojis in the questionnaire [30].
6. At the beginning of the test, consumers need to be instructed about the difference between emotions (brief, intense and often focused on a referent [13]), and mood (low intensity and long duration feelings [15]).
7. Evoking the consumption context may be done, whether by asking explicitly for it to the consumers, or by asking questions to the consumers directly related to the typical consumption context of the product [28].
8. The reader interested in this kind of effect could find useful and suitable references in the bibliography [31–36].

References

1. Talavera M, Chambers E IV (2017) Using sensory sciences help products succeed. Br Food J 119(10):2130–2144
2. Kemp SE, Hollowood T, Hort J (2011) Sensory evaluation: a practical handbook. Wiley, Oxford
3. Lawless HT, Heymann H (2010) Sensory evaluation of food: principles and practices. Springer, New York
4. Gutjar S, de Graaf C, Kooijman V et al (2015) The role of emotions in food choice and liking. Food Res Int 76:216–223
5. Ng M, Chaya C, Hort J (2013) Beyond liking: comparing the measurement of emotional response using EsSense profile and consumer defined check-all-that-apply methodologies. Food Qual Prefer 28(1):193–205
6. Silva AP, Jager G, Voss HP et al (2017) What's in a name? The effect of congruent and incongruent product names on liking and emotions when consuming beer or non-alcoholic beer in a bar. Food Qual Prefer 55:58–66
7. Damasio AR (2006) Descartes' error. Random House
8. Edwards JS, Hartwell HJ, Brown L (2013) The relationship between emotions, food consumption and meal acceptability when eating out of the home. Food Qual Prefer 30(1): 22–32
9. Ferrarini R, Carbognin C, Casarotti EM et al (2010) The emotional response to wine consumption. Food Qual Prefer 21(7):720–725
10. Ng M, Chaya C, Hort J (2013) The influence of sensory and packaging cues on both liking and emotional, abstract and functional conceptualisations. Food Qual Prefer 29(2):146–156
11. Torrico DD, Fuentes S, Viejo CG et al (2018) Images and chocolate stimuli affect physiological and affective responses of consumers: a cross-cultural study. Food Qual Prefer 65:60–71
12. Ng M, Hort J (2015) Insights into measuring emotional response in sensory and consumer research. In: Rapid sensory profiling techniques. Woodhead Publishing, Duxford
13. King SC, Meiselman HL (2010) Development of a method to measure consumer emotions associated with foods. Food Qual Prefer 21(2):168–177

14. Sander D (2013) Models of emotion: the affective neuroscience approach. University Press, Cambridge
15. Coppin G, Sander D (2021) Theoretical approaches to emotion and its measurement. In: Emotion measurement. Woodhead Publishing, Duxford
16. Van Zyl H, Chaya C (2021) Chapter 23. Emotion in beverages. In: Meiselman HL (ed) Emotion measurement. Elsevier-Woodhead Publishing, Duxford
17. Dorado R, Pérez-Hugalde C, Picard A et al (2016) Influence of first position effect on emotional response. Food Qual Prefer 49: 189–196
18. Wakeling IN, MacFie HJ (1995) Designing consumer trials balanced for first and higher orders of carry-over effect when only a subset of k samples from t may be tested. Food Qual Prefer 6(4):299–308
19. Bradley MM, Lang PJ (1994) Measuring emotion: the self-assessment manikin and the semantic differential. J Behav Ther Exp Psychiatry 25(1):49–59
20. Desmet PMA, Hekkert P, Jacobs JJ (2000) When a car makes you smile: development and application of an instrument to measure product emotions. Adv Consum Res 27(1): 111–117
21. Rizo A, Bartu A, Laguna L et al (2023) Effect of an opinion app on expectations and emotional responses of young consumers toward white wines. Food Qual Prefer 103:104706
22. Thomson DM, Crocker C (2013) A data-driven classification of feelings. Food Qual Prefer 27(2):137–152
23. Gmuer A, Guth JN, Runte M, Siegrist M (2015) From emotion to language: application of a systematic, linguistic-based approach to design a food-associated emotion lexicon. Food Qual Prefer 40:77–86
24. Nestrud MA, Meiselman HL, King SC, Lesher LL, Cardello AV (2016) Development of EsSense25, a shorter version of the EsSense profile®. Food Qual Prefer 48:107–117
25. Danner L, Ristic R, Johnson TE, Meiselman HL, Hoek AC, Jeffery DW, Bastian SE (2016) Context and wine quality effects on consumers' mood, emotions, liking and willingness to pay for Australian shiraz wines. Food Res Int 89: 254–265
26. Silva AP, Jager G, Van Bommel R et al (2016) Functional or emotional? How Dutch and Portuguese conceptualise beer, wine and non-alcoholic beer consumption. Food Qual Prefer 49:54–65
27. Mora M, Dupas de Matos A, Fernández-Ruiz V, Briz T, Chaya C (2020) Comparison of methods to develop an emotional lexicon of wine: conventional vs rapid-method approach. Food Qual Prefer 83:103920
28. Dorado R, Chaya C, Tarrega A et al (2016) The impact of using a written scenario when measuring emotional response to beer. Food Qual Prefer 50:38–47
29. Chaya C (2022) Chapter 2: Sensometría. In: Pérez-Elortondo FJ, Salvador-Moya MD (eds) Análisis sensorial de alimentos y respuesta del consumidor. Asociación Española de Profesionales del Análisis Sensorial, Editorial Acribia, Zaragoza
30. King SC, Meiselman HL, Thomas CB (2013) Measuring emotions associated with foods: important elements of questionnaire and test design. Food Qual Prefer 28(1):8–16
31. van Zyl H, Meiselman HL (2015) The roles of culture and language in designing emotion lists: comparing the same language in different English and Spanish speaking countries. Food Qual Prefer 41:201–213
32. Chaya C, Pacoud J, Ng M et al (2015) Measuring the emotional response to beer and the relative impact of sensory and packaging cues. J Am Soc Brew Chem 73(1):49–60
33. Yang Q, Dorado R, Chaya C et al (2018) The impact of PROP and thermal taster status on the emotional response to beer. Food Qual Prefer 68:420–430
34. Mora M, Urdaneta E, Chaya C (2018) Emotional response to wine: sensory properties, age and gender as drivers of consumers' preferences. Food Qual Prefer 66:19–28
35. Mora M, Urdaneta E, Chaya C (2019) Effect of personality on the emotional response elicited by wines. Food Qual Prefer 76:39–46
36. Mora M, Dupas De Matos A, Vázquez-Araújo L et al (2021) Exploring young consumers' attitudes and emotions to sensory and physicochemical properties of different red wines. Food Res Int 143:110303

Part III

Instrumental and Sensory Procedures

Chapter 17

Analysis of Wine Impact Odorants by Gas Chromatography-Olfactometry

Ozlem Kilic-Buyukkurt, Gamze Guclu, Onur Sevindik, Hasim Kelebek, and Serkan Selli

Abstract

Aroma is a primary determinant of wine quality, consumer acceptance, and preference. More than one thousand aroma compounds may be responsible for the aroma of wines. Despite the fact that many aromatic compounds are available in wines, the application of gas chromatography-olfactometry (GC-O) in flavor analysis is considered to be a valuable technique to characterize odor-active compounds responsible for the characteristic odor of a wine sample. GC-O is based on the use of human assessors, a sensitive detector, to detect and evaluate aroma compounds from a chromatographic column eluate. There are various GC-O methods based on different principles (dilution analysis, detection frequency, and direct intensity) to obtain data on aroma-active compounds in GC-O. This chapter explains in detail the application of dilution analysis, detection frequency, and direct intensity methods from GC-O techniques in conjunction with solvent-assisted flavor evaporation (SAFE) to elucidate key odorants of wines.

Key words Gas chromatography-olfactometry, Wine, Aroma-active, Key odorants, Dilution analysis, Detection frequency, Direct intensity

1 Introduction

One of the most crucial characteristics of wine is its flavor which significantly influences its quality characteristics and consumer preference. The flavor of wines is formed as a result of the interaction of hundreds of different aroma compounds [1]. The contribution of each compound to the aroma profile is different. The individual importance of each compound on the final wine aroma depends on the correlation between chemical composition and odor perception thresholds because most of the volatiles are present at concentrations near or below their individual odor sensory thresholds [2]. These compounds in wines are called aroma-active compounds or key odorants [3]. It is important for aroma analysis to identify these aroma-active compounds that are responsible for the overall

María Ángeles Pozo-Bayón and Carolina Muñoz González (eds.), *Wine Analysis and Testing Techniques*, Methods and Protocols in Food Science, https://doi.org/10.1007/978-1-0716-3650-3_17,

aroma of the food and distinguish them from other volatiles [4]. The main step in the study of key odorant compounds is selecting a suitable extraction method that gives an aromatic extract as similar as possible to the studied wine sample. Various extraction techniques have been studied for the volatile compounds in wines, including liquid-liquid extraction (LLE), solvent-assisted flavor evaporation (SAFE), solid-phase microextraction (SPME), stir bar sorptive extraction (SBSE), solid-phase dynamic extraction (SPDE), headspace techniques (HS), and solid-phase extraction (SPE) [1, 5–7].

GC or GC-MS with olfactometry (GC-O or GC-MS-O) is a commonly used combined method for the determination of aroma-active compounds in wines [8, 9]. In this method, gas chromatography is combined with the olfactometer which uses the human nose as a detector [10–12], and this method was first proposed by Fuller et al. [13]. Especially in the GC-MS-O system, the olfactory (sniffing) port is connected in parallel to the flame ionization detection (FID) or a mass spectrometer (MS) [14, 15]. In this way, the aroma compounds in the extracts given GC-MS-O are sent to FID and MS simultaneously, allowing the identification of the aroma compound and its detection by the human nose by sniffing (Fig. 1). There are various methods based on different principles of obtaining data on aroma-active compounds in GC-O. These methods are divided into three categories: dilution analysis, detection frequency, and direct intensity methods [8, 16–18]. The use of olfactometric methods enables the determination of the characteristic aroma notes of many compounds found in the volatile fractions of wines. Also, these methods are used to identify the new aroma-active compounds in wines.

The aim of the present chapter is to explain technically and step-by-step the GC-O analysis of an aroma extract from wine by using solvent-assisted aroma extraction (SAFE) method and aroma extract dilution analysis (AEDA), detection frequency, and direct intensity methods allowing the determination of wine aroma-active compounds.

2 Materials

- The determination of aroma-active compounds by SAFE and AEDA procedures was explained according to a previously detailed paper published by Issa-Issa et al. [19].
- The procedure mainly consisted of a SAFE unit (Glasblaserei Bahr, Manching, Germany) equipped with a vacuum pump (10^3 Pa; Vacuubrand DCP 3000, Wertheim, Germany) for wine volatiles extraction (Fig. 2).
- A Gas chromatography (GC) instrument (Agilent 6890) equipped with a mass selective detector (MSD) (5973 Network

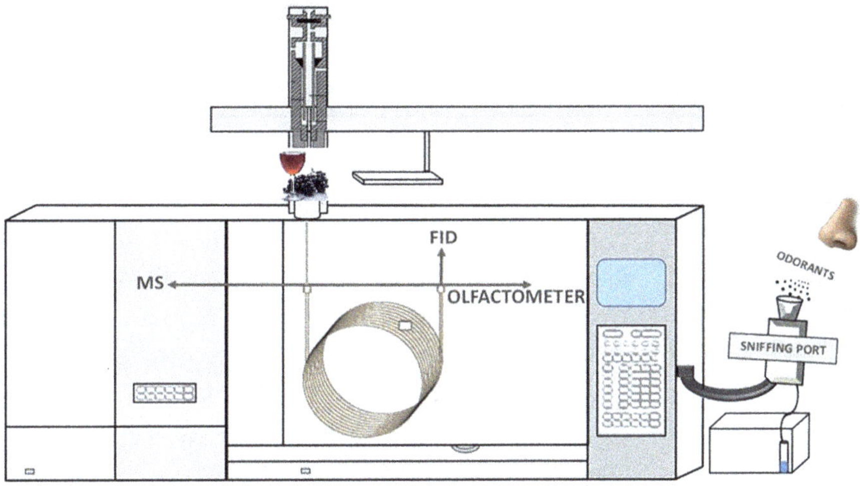

Fig. 1 Gas chromatography-mass spectrometry-olfactometry system (GC-MS-O)

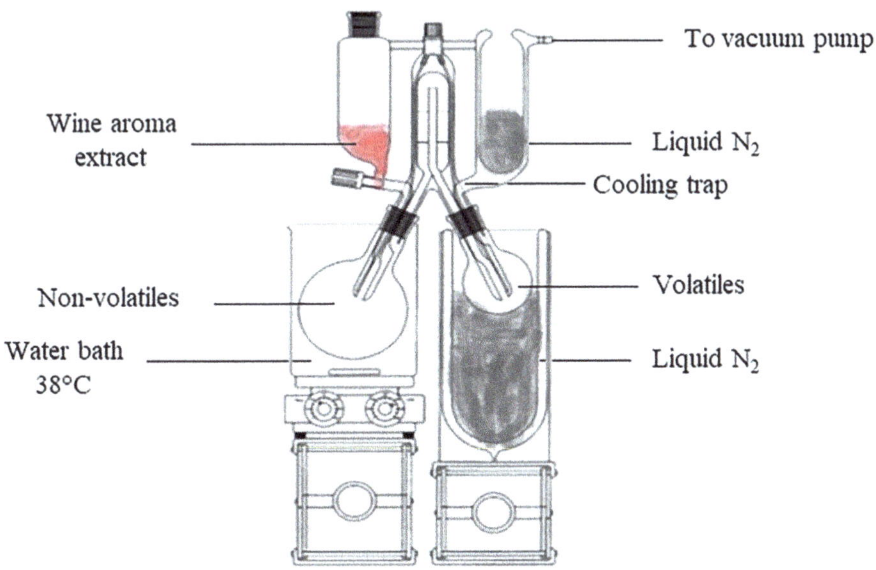

Fig. 2 Solvent-assisted flavor extraction (SAFE) unit

159 MSD, Agilent Technologies, Wilmington, Delaware, USA), flame ionization detector (FID), and a Gerstel ODP-2 olfactometry sniffing device (Baltimore, Maryland, USA) along with humidified air at 40 °C using deactivated fused silica capillary column (30 cm × 0.3 mm) (*see* **Notes 1** and **2**).

- A well-known column type for volatile separations, DB-Wax column (30 m length × 0.25 mm i.d.× 0.5 μm thickness, J&W Scientific, Folsom, California, USA), used for the elucidation of each aroma compound and helium is used as carrier gas. Separated volatile compounds were identified with an MS detector while quantified with an FID using a representative internal standard.

3 Methods

3.1 Extraction of Wine Aroma Compounds by Solvent-Assisted Flavor Evaporation (SAFE)

1. Mix 100 mL of wine, 5 mL of 4-nonanol as the internal standard, and 100 mL of dichloromethane into a 500 mL glass flask and stir at 4 °C for 60 min under an N_2 atmosphere.
2. Centrifuge the mixed samples at 4 °C, 5500 rpm for 15 min.
3. Place the organic phase (solvent) into the dropping funnel of the transfer head and separate the mixture (*see* **Note 3**).
4. Control the temperature with a thermostat to maintain SAFE system in cold water at 4 °C to ensure a continuous temperature throughout distillation and to avoid condensation of the volatile compounds.
5. Collect the aroma by condensation of compounds passed into a receiving vessel using a cooling trap (*see* **Note 4**).
6. After separation, remove the receiving vessel and allow it to thaw out at room temperature for 30 min.
7. Dehydrate the extract with anhydrous sodium sulfate, and concentrate the aromatic extract to 5 mL in a Kuderna Danish concentrator (Sigma Aldrich, St. Louis, MO) fitted with a Snyder column (Supelco, St. Quentin, France) to evaporate the solvent by increasing the surface area as well as for the collection without any loss of volatile compounds in a balloon and then to 200 μL under a gentle stream of purified nitrogen.
8. Store the extract at −20 °C in a 2 mL glass vial fitted with a Teflon-lined cap before the analysis.

3.2 GC–FID, GC–MS, and GC–O Analyses of Wine Aroma Compounds

1. Increase gradually the oven temperature of the column as follows:

   (1) initially hold the column at 40 °C for 10 min, (2) increase from 40 to 160 °C at a rate of 3 °C/min, (3) subsequently increase the temperature to 240 °C at 6 °C/min rate, (4) and finally hold it at 240 °C for 25 min.
2. Set up the column pressure at a constant value of 20.0 psi and injection volume at 3 μL in pulsed splitless mode.
3. Set the injector and FID detectors at 270 °C and 280 °C, respectively.
4. The MS (electronic impact ionization) conditions are the ionization energy of 70 eV, mass range m/z of 33–300 amu at 2.0 scan/s.

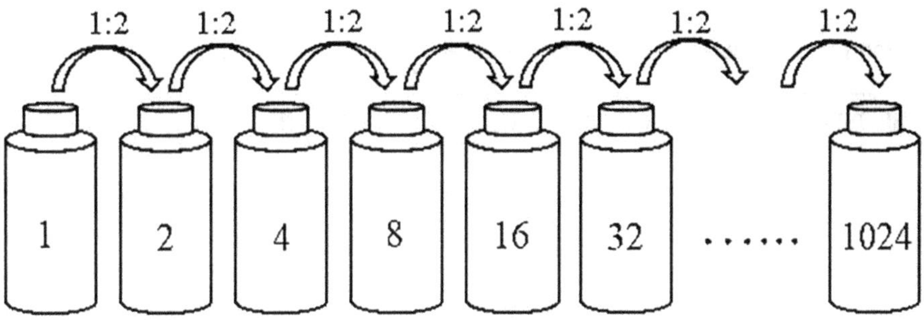

Fig. 3 Stepwise wine aroma extract dilution

5. Determine the concentrations of aroma compounds by the application of the internal standard method according to the following equation (Eq. 1) and express the results as μg/L resulting from the relative peak area to the internal standard (*see* **Note 5**).

$$C_i = (A_i/A_{st}) \times C_{st} \times \mathrm{RF} \times \mathrm{CF} \tag{1}$$

   C_i: Concentration of the aroma compound
   A_i: The peak area of the aroma compound
   A_{st}: Peak area of the internal standard (4-nonanol)
   C_{st}: Concentration of internal standard (40 μg/L)
   RF: Response factor (*see* **Note 6**)
   CF: Calculation factor (factor for converting sample amount to liters (for the unit conversion): 10)

6. Determine all compounds by comparing their retention indexes and mass spectra on the DB-Wax column (*see* **Note 7**).

3.3 Aroma Extract Dilution Analysis (AEDA)

1. Apply the AEDA developed by Schieberle and Grosch [20] with at least 2 experienced panelists (*see* **Note 8**) to determine the impact of individual aroma active compounds on the overall wine aroma.
2. Dilute gradually the concentrated aroma extract of the wine sample stepwise with dichloromethane (1:2, v/v) as shown in Fig. 3 (*see* **Note 9**).
3. Inject each diluted sample (3 μL) into the GC-MS-O and sniff the effluent with at least 2 experienced panelists (*see* **Notes 10–12**).
4. Identify the aroma-active compounds by the flavor dilution (FD) factor of the last dilution. Increasing the FD value indicates the activity level of the aroma compound. For instance, Issa-Issa et al. [19] investigated the aroma-active compounds of *Fondillón* wine using the AEDA method. Some of the aroma-active compounds determined according to the results of this study are given in Table 1. Accordingly, diethyl glutarate

Table 1
Flavor dilution factor calculation in aroma extract dilution analysis [19]

Compounds/Dilutions	4	8	16	32	64	128	256	512	1024	FD factor
Ethyl propanoate	✓	✓	✓	✓	✓	□	□	□	□	64
Ethyl lactate	✓	✓	✓	✓	✓	✓	✓	✓	□	512
Diethyl glutarate	✓	□	□	□	□	□	□	□	□	4
Diethyl DL malate	✓	✓	✓	✓	✓	✓	□	□	□	128
2,3-Butanediol	✓	✓	□	□	□	□	□	□	□	8
Phenylethyl alcohol	✓	✓	✓	✓	✓	✓	✓	✓	✓	1024
Furfural	✓	✓	✓	□	□	□	□	□	□	16
γ-Butyrolactone	✓	✓	✓	✓	□	□	□	□	□	32
Whiskey lactone	✓	✓	✓	□	□	□	□	□	□	16

was determined with the lowest flavor dilution factor (FD: 4), while phenylethyl alcohol had the highest flavor dilution factor (FD: 1024). In other words, phenylethyl alcohol is the compound that contributes the most to the overall aroma of *Fondillón* wine followed by ethyl lactate (FD: 512).

3.4 Detection Frequency Method (DFM)

1. Perform the sniffing by 6–12 panelists with no training on the same aroma extract.
2. Determine the intensity of an aroma compound according to the number of panelists that detect it.
3. The number of panelists indicating the same odor at the same retention time during the sniffing of the extract is expressed as the detection frequency [11]. The aroma-active compound with a high detection frequency is thought to be proportionally high, and this is associated with the intensity of the aroma compound (*see* **Note 13**).
4. Quantify the results for each odor using so-called olfactometric indices, such as NIF (Nasal Impact Frequency) – SNIF (Surface of Nasal Impact Frequency) values, and modified frequency (MF) (*see* **Note 14**).

3.5 Direct-Intensity Method (DIM)

The direct intensity method includes a single time-averaged measurement (posterior intensity) and a dynamic measurement that continuously records the onset, maximum intensity, and decline of the eluting odor (time intensity).

3.5.1 Time-Intensity Method (TIM)

1. Apply the sniffing process without preparing dilutions of aroma extracts using an electronic scale (*see* **Note 15**).

2. Perform the odor-specific magnitude estimation (OSME) analysis by four experienced panelists and each extract is sniffed four times (*see* **Note 16**).
3. Measure the intensity of the odor using a horizontal slide bar and a 16-point structured scale with a range of 0 (none) to 15 (strong).
4. Record continuously the existence of an odor, maximum intensity, and decline of an odor.
5. Create an aromagram which is called Osmegram, using the aroma intensities and perception times given by the panelists (*see* **Note 17**).

3.5.2 Posterior Intensity Method

1. Perform the sniffing by 3–10 trained panelists (*see* **Note 8**) to measure the intensity. Panelists only rate this maximum odor intensity after the compound has been eluted from the GC column [21].
2. Rate the perceived odor intensity of each aroma compound in a memorized five-point intensity interval scale after a peak has eluted from the olfactory detection port.

A previous study is explained to understand better the two methods (detection frequency and direct intensity) mentioned above. Botelho et al. [17] applied detection frequency and posterior intensity methods to determine the aroma-active compounds of Aragonez clonal wines. A panel of 8 panelists was used to implement these methods. In the detection frequency method, panelists sniffed the column's effluent and pressed a joystick (electronic scale) when they detected an odor. In the posterior intensity method, previously trained panelists used a memorized five-point intensity interval scale (1: very mild; 2: mild; 3: moderate; 4: strong; 5: very strong). Some aroma-active compounds determined according to the results of this study are given in Table 2. The fact that the number of panelists is 8 in the detection frequency method means that 8 out of 8 panelists perceive that odor. In the posterior intensity method, it refers to the average of the scores on a 5-point scale. Ethyl isobutyrate, 3-methylbutanoic acid, and 2,5-dimethyl-4-hydroxy-3(2H)-furanone compounds were also identified by 8 panelists. However, the intensity of these compounds varied between 2.4 and 4 according to the posterior method. Also, in the detection frequency method, the odors detected by less than three panelists are considered odor noise and eliminated such as ethyl octanoate and 2-ethyl-4-hydroxy-5-methyl-3(2H)-furanone as seen in Table 2.

Table 2
Some of the aroma-active compounds determined according to detection frequency and posterior intensity methods [17]

Compounds	Odor descriptions	Number of panelists	Intensity
Ethyl isobutyrate	Fruity	8	2.4
Ethyl 2-methylbutanoate	Fruity	7	1.1
Ethyl octanoate	Fruity, floral	3	0.8
Benzaldehyde	Plastic	6	1.6
γ-Butyrolactone	Smoky, hot	4	0.6
β-Damascenone	Floral, fruity	7	2.4
3-Methylbutanoic acid	Stinky, cheese	8	4.0
2,5-Dimethyl-4-hydroxy-3(2 H)-furanone	Burnt sugar, candy cotton	8	3.9
2-Ethyl-4-hydroxy-5-methyl-3(2H)-furanone	Burnt sugar, candy cotton	3	1.0

4 Notes

1. The laboratory where the GC-O is located must be free from all foreign odors and sounds. During the analysis, soundproof headphones can be used to avoid distraction from noise. The temperature in the laboratory should also be controlled for the panelist's comfort as well as the instrument.
2. The GC system divides the effluent into 3 equal flows (1:1:1) to the MS, FID, and sniffing port to identify and quantify the aroma and aroma-active compounds of the wine simultaneously.
3. The mixture is separated in the distillation vessel (10 mL/min) partially submerged in a warm water bath of 38 °C.
4. There is liquid nitrogen in the cooling trap and the aroma compounds condense and freeze due to the sudden temperature drop there.
5. The concentration equivalent of the 4-nonanol, internal standard, is 40 μg/L.
6. The internal standard, whose concentration is known, is added to both the water and the wine sample, and the same aroma extraction method is performed for both, and it is checked whether there is any loss in the concentration of the internal standard.
7. The commercial spectra databases (Wiley 10, Flavor 2, NBS 75 k) and the internal library of the instrument created in

earlier studies is compared to the retention indexes and mass spectra on the DB-Wax column to identify all aroma compounds. Retention indices are calculated based on retention times of the standard linear n-alkane series (C_6-C_{32}) from the following formula:

$$RI = 100\ N + 100n\ \left(t_{R,a} - t_{R,N}\right) / \left(t_{R(N+n)} - t_{R,N}\right)$$

N: the carbon number of the lower alkane

n: the difference in carbon number between two n-alkanes that bracket the analyte

$t_{R,a}$: the retention time of the unknown component

$t_{R,N}$: the retention time of the alkane of lower carbon numbers

$t_{R(N+n)}$: the retention time of the alkane of upper carbon numbers [22].

8. The training of panelists includes learning terminology using standard compounds, using an assessment scale, determining compounds and their concentrations in the reference mixture with olfactometry, and sniffing the studied sample [23].
9. First, mix the 100 μL concentrated extract and 100 μL dichloromethane. This represents your 1:1 diluted sample. After, mix the 100 μL of 1:1 diluted sample and 100 μL dichloromethane (1:2 diluted sample). Again, mix the 100 μL of 1:2 diluted sample and 100 μL dichloromethane (1:4 diluted sample). Keep diluting the sample until no odor is detected.
10. The longer the GC-O sniffing time, the more fatigued panelists become, which can affect performance. Therefore, if the chromatographic program is longer than 25 min, the entire program should be divided into several parts and each part should be sniffed by a different panelist.
11. Panelists should not smoke, consume strongly flavored foods, and use perfumes or strong deodorants for at least 1 h before the GC-O analysis of wine [21].
12. First, a 1:1 diluted sample is sniffed. As the panelist sniffs identify the greenish, vinegary, earthy, etc. odors of the compounds. This is called identification sniffing. Then, as the other diluted samples are sniffed in order, it is checked whether the compounds whose smell is identified are present in that dilution and are marked in the list. It does not require specific software and panelists do not assign a value for the intensity or duration of the odor.
13. The number of panelists detecting an aroma-active compound at the olfactory detection port is used as an estimate of the odor's intensity. The odors detected by less than three panelists are considered odor noise [17].

14. In the NIF-SNIF technique, the peak height indicates the percentage of panelists who perceived the aroma-active compound and this is expressed as the NIF (%). The NIF value, retention time, and odor duration of each compound are plotted similarly to a chromatogram. The peak area is stated as the SNIF value [5]. The duration of the odor is measured by the panelists and SNIF is obtained by multiplying the detection frequency by the average duration according to the following equation [12].

$$\mathrm{SNIF} = \%\mathrm{Frequency} \times \mathrm{Duration}$$

The modified frequency (MF) technique takes into account the odor's intensity in accordance with a scale and odor duration is not taken into consideration. MF (%) is obtained by multiplying the detection frequency (DF) by the average intensity (expressed as a percentage of the maximum intensity, I) as seen in the following equation [18].

$$\mathrm{MF}\ (\%) = [\mathrm{DF}\ (\%) \times \mathrm{I}\ (\%)]^{1/2}$$

15. During sniffing, the panelists determine the intensities of the aroma compounds with the help of a special electronic scale and also, they define the quality definitions and the time when the odor is noticed [11, 24].
16. The odor-specific magnitude estimation (OSME) is a dynamic method used to measure the perceived odor intensity of a compound eluting from the GC-O [7, 25].
17. Osmegram represents odor intensity as a function of an analyte's retention time. The height of the peak expresses the maximum odor intensity of an analyte, while the width corresponds to the odor duration [14]. The area under the odor intensity peak together with the maximum odor intensity of the compounds represents the correlation of the concentration of the compounds in the GC effluent [10].

References

1. Selli S, Cabaroglu T, Canbas A, Erten H, Nurgel C, Lepoutre JP, Gunata Z (2004) Volatile composition of red wine from cv. Kalecik karası grown in central Anatolia. Food Chem 85(2):207–213
2. Villamor RR, Ross CF (2013) Wine matrix compounds affect perception of wine aromas. Annu Rev Food Sci Technol 4:1–20
3. Selli S, Canbas A, Varlet V, Kelebek H, Prost C, Serot T (2008) Characterization of the most odor-active volatiles of orange wine made from a Turkish cv. Kozan (*Citrus sinensis* L. Osbeck). J Agric Food Chem 56(1):227–234
4. Grosch W (1993) Detection of potent odorants in foods by aroma extract dilution analysis. Trends Food Sci Technol 4(3):68–73
5. d'Acampora Zellner B, Dugo P, Dugo G, Mondello L (2008) Gas chromatography–olfactometry in food flavor analysis. J Chromatogr A 1186:123–143
6. Andujar-Ortiz I, Moreno-Arribas MV, Martín-Álvarez PJ, Pozo-Bayón MA (2009) Analytical performance of three commonly used extraction methods for the gas chromatography–mass spectrometry analysis of wine volatile

compounds. J Chromatogr A 1216:7351–7357

7. Muñoz-González C, Rodríguez-Bençomo JJ, Moreno-Arribas MV, Pozo-Bayón MÁ (2011) Beyond the characterization of wine aroma compounds: looking for analytical approaches in trying to understand aroma perception during wine consumption. Anal Bioanal Chem 401:1497–1512
8. Serot T, Prost C, Visan L, Burcea M (2001) Identification of the main odor-active compounds in musts from French and Romanian hybrids by three olfactometric methods. J Agric Food Chem 49:1909–1914
9. Brattoli M, Cisternino E, Dambruoso PR, de Gennaro G, Giungato P, Mazzone A, Palmisani J, Tutino M (2013) Gas chromatography analysis with olfactometric detection (GC-O) as a useful methodology for chemical characterization of odorous compounds. Sensors 13:16759–16800
10. van Ruth SM (2001) Methods for gas chromatography-olfactometry: a review. Biomol Eng 17:121–128
11. Falcão LD, de Revel G, Rosier JP, Bordignon-Luiz MT (2008) Aroma impact components of Brazilian Cabernet Sauvignon wines using detection frequency analysis (GC-olfactometry). Food Chem 107:497–505
12. de-la-Fuente-Blanco A, Ferreira V (2020) Gas chromatography olfactometry (GC-O) for the (Semi) quantitative screening of wine aroma. Foods 9:1892
13. Fuller GH, Steltenkamp R, Tisserend GA (1964) The chromatography with human sensor: perfumer model. Ann N Y Acad Sci 116: 711–724
14. Plutowska B, Wardencki W (2008) Application of gas chromatography–olfactometry (GC–O) in analysis and quality assessment of alcoholic beverages – a review. Food Chem 107:449–463
15. Guclu G, Kelebek H, Selli S (2020) Chapter 6 – Saffron (*Crocus sativus* L.): its aroma and key odorants. In: Sarwat M, Sumaiya S (eds) Saffron. Academic Press, pp 69–82
16. Ferreira V, Petka J, Aznar M (2002) Aroma extract dilution analysis. Precision and optimal experimental design. J Agric Food Chem 50: 1508–1514
17. Botelho G, Caldeira I, Mendes-Faia A, Clímaco MC (2007) Evaluation of two quantitative gas chromatography–olfactometry methods for clonal red wines differentiation. Flavour Fragr J 22:414–420
18. Welke EJ, Hernandes KC, Nicolli KP, Barbará JA, Telles Biasoto AC, Zini CA (2021) Role of gas chromatography and olfactometry to understand the wine aroma: achievements denoted by multidimensional analysis. J Sep Sci 44:135–168
19. Issa-Issa H, Guclu G, Noguera-Artiaga L, López-Lluch D, Poveda R, Kelebek H, Selli S, Carbonell-Barrachina ÁA (2020) Aroma-active compounds, sensory profile, and phenolic composition of *Fondillón*. Food Chem 316: 126353
20. Schieberle P, Grosch W (1987) Evaluation of the flavour of wheat and rye bread crusts by aroma extract dilution analysis. Z Lebensm Unters Forsch 185:111–113
21. Delahunty CM, Eyres G, Dufour JP (2006) Gas chromatography-olfactometry. J Sep Sci 29:2107–2125
22. van den Dool H, Kratz PD (1963) A generalization of the retention index system including linear temperature programmed gas liquid partition chromatography. J Chromatogr 11:46
23. Vene K, Seisonen S, Koppel K, Leitner E, Paalme T (2013) A method for GC–olfactometry panel training. Chemosens Percept 6:179–189
24. Pollien P, Ott A, Montigon F, Baumgartner M, Munoz-Box R, Chaintreau A (1997) Hyphenated headspace-gas chromatography-sniffing technique: screening of impact odorants and quantitative aromagram comparisons. J Agric Food Chem 45:2630–2637
25. Miranda-Lopez R, Libbey LM, Watson BT, McDaniel M (1992) Odor analysis of Pinot Noir wines from grapes of different maturities by a gas chromatography-olfactometry technique (Osme). J Food Sci 57(4):985–993

INDEX

Maria Angeles del Pozo Bayon and Carolina Muñoz González (eds.), *Wine Analysis and Testing Techniques*, Methods and Protocols in Food Science , https://doi.org/10.1007/978-1-0716-3650-3,

Printed in the USA
CPSIA information can be obtained
at www.ICGtesting.com
LVHW080832010224
770330LV00005B/490